NASA SP-275

I0473950

MONITORING EARTH RESOURCES FROM AIRCRAFT AND SPACECRAFT

By

ROBERT N. COLWELL

and other personnel of the

Forestry Remote Sensing Laboratory
School of Forestry and Conservation
University of California

Scientific and Technical Information Office   1971
NATIONAL AERONAUTICS AND SPACE ADMINISTRATION
Washington, D.C.

For sale by the Superintendent of Documents,
U.S. Government Printing Office, Washington, D.C. 20402
Stock Number 3300–0406
Price $4.00
Library of Congress Catalog Card No. 73–170325

Preface

Man soon will be confronted by one of the most serious crises of his existence. Basically, this crisis is developing because the world's population is rapidly increasing at the very time when many of its natural resources have dwindled to a very low level. The supply-versus-demand problem is made even more serious by another recent development: Within the last decade ~~a greatly increased awareness of the have-nots, both in the United States and elsewhere, combined with demands of the affluent for an ever higher standard of living, has resulted in~~ a tremendous increase in the per capita demand for Earth resources at virtually every economic level.

The Food and Agriculture Organization of the United Nations has stated that the production of food will have to be doubled by 1980 and trebled by 2000 to provide a decent level of nutrition to the world's people. In recent testimony given to the U.S. House of Representatives by personnel of the lumber and housing industries, there was a consensus that 26 million homes must be built in the United States during the next decade. This represents a construction rate nearly triple that at present, with a proportionate increase in the demand for lumber and other materials. Experts from the U.S. Department of Interior have repeatedly warned that the reserves of copper, lead, zinc, and many other important nonrenewable mineral resources will be exhausted within the next two or three decades at the estimated rate of consumption.

In view of such circumstances, we need the wisest possible management of the Earth's resources. An important first step leading to such management is that of obtaining accurate resource inventories, quickly and at frequent intervals.

This report first describes an experiment that sought to determine the extent to which Earth resources might be monitored by means of periodic inventories made with the aid of aerial and space photography. It then presents the results obtained from that experiment in each of several geographic test areas. Special emphasis is given to *vegetation* resources for two reasons: (1) their intelligent management requires that they be monitored at frequent intervals, and (2) the economic benefits derivable from such monitoring of vegetation resources are potentially very great. Following the chapters in which the results obtained in various geographic areas have been reported, consideration is given to the potential value of such results to the resource manager. The final chapter summarizes the experimental results, both quantitatively and qualitatively. Then, on the basis of these results, conclusions are drawn as to the advantages, limitations, and overall feasibility of monitoring Earth resources with the aid of aerial and space photography.

The conduct of research leading to the preparation of this report was made possible through the cooperation of a large number of individuals and institutions. The timely support of Leonard Jaffe and his staff at NASA Headquarters in Washington, D.C., made it possible to take advantage, on very short notice, of an unusual opportunity—that of incorporating the S065 multispectral terrain experiment into the busy schedule of Apollo 9 activities. Allen L. Grandfield, as project engineer for the experiment, was completely successful in obtaining the necessary state of readiness of a four-camera system and in briefing the Apollo 9 astronauts on the use of this complex experimental equipment. The three astronauts who took the space photography were James McDivitt, Dave Scott, and Russ Schweickart. Dr. John Dornbach, Assistant Chief for Programs at the NASA Manned Spacecraft Center's Earth Resources Division, was project scientist and Dr. Paul Lowman, of the NASA Goddard Space Flight Center, was the principal investigating scientist. The great success of the Apollo 9 flight, from the photographic standpoint, is attributable in large measure to the organizational skills of these scientists and to the wisdom of their "real time" decisions during the flight as to areas that might best be photographed. The information on orbital path, time over target, and problems of cloud cover was provided by a highly motivated and indefatigable group of personnel of the NASA Earth Resources Survey program at the Manned Spacecraft Center in Houston, under the direction of Dr. Robert O. Piland, Acting Chief of the Earth Resources Survey program. Three coinvestigators for the experiment (Dr. Phil Slater of the University of Arizona, Dr. Edward F. Yost of Long Island University, and Herbert A. Tiedemann of the Manned Spacecraft Center) also provided valuable information for our report. Personnel of the Infrared and Optical Sensor Laboratory of the University of Michigan and of Bendix Corp. at Ann Arbor, Mich., were highly successful in obtaining supporting data with optical mechanical scanners at the time of the Apollo 9 overflights. Ground truth provided for the Imperial Valley area by Harold Greene and other U.S. Department of Agriculture (USDA) personnel was ably summarized by Norma Spansail of the University of Michigan and made available for our use. As provided in our NASA contract, we were able to make funds available to Dr. Charles Poulton and his associates at Oregon State University to study Apollo 9 photography of range resources in southern Arizona. Their excellent contribution appears as chapter 5 of this report.

Special acknowledgment is given to Lawrence R. Pettinger for the material he contributed from his special report entitled "Analysis of Earth Resources on Sequential High Altitude Multiband Photography," and for the effort he made integrating this material with findings from the study of Apollo 9 space photography.

The valuable assistance of these individuals and organizations, together with the contributions of others too numerous to mention, is hereby gratefully acknowledged.

The work dealt with in this report was funded under the NASA Earth Resources Survey program. Certain of these funds were provided to the Forestry Remote Sensing Laboratory at the University of California for the specific purpose of analyzing multiband photography that was taken by the Apollo 9 astronauts. The remaining funds were provided under the NASA–USDA Remote Sensing Research Program and were used to finance work at the Forestry Remote Sensing Laboratory, the Department of Range Management at Oregon State University, the Pacific Southwest Forest and Range Experiment Station, and the Rocky Mountain Forest and Range Experiment Station. The latter two stations are research facilities of the U.S. Forest Service.

ROBERT N. COLWELL,
Associate Director, Space Sciences Laboratory
Coordinator, NASA–USDA Forestry
Remote Sensing Research Program

BERKELEY, CALIF.

Contents

Introduction

ROBERT N. COLWELL

There are several reasons why Earth resource surveys can best be made through the use of aircraft and spacecraft.

The first of these is clearly implied in the simple statement that "the face of the land looks to the sky." The task of inventorying Earth's resources is, first of all, one of delineating boundaries between one resource characteristic and another. When confined to Earth, man often has great difficulty in recognizing and delineating these boundaries.

A second reason for using aircraft or spacecraft is that the broad synoptic view, so essential for the quick and economical delineation of Earth resource features, can only be obtained in this way. Vast areas can be viewed at a single point in both space and time and hence under relatively uniform lighting conditions.

The ability of the aircraft or spacecraft to travel quickly from one camera station to another is a related advantage of great importance. For example, an Earth-orbiting spacecraft can be placed in a Sun-synchronous orbit such that all portions of the illuminated hemisphere can be photographed at the same local Sun time. This capability, which is to be incorporated in future Earth Resource Technology satellites, permits even larger areas to be photographed under nearly uniform lighting conditions, thereby facilitating the photoidentification of Earth resource features.

Perhaps another advantage of the view obtained from air or space should be mentioned, even though it is somewhat corollary to the advantages already mentioned. Without the ability to view subtle lineaments and other patterns simultaneously over a vast areal extent, the resource analyst might never become aware of their existence. Often it is only by his being able to discern each part of the pattern in relation to all other parts that he is able to discern the feature and thereby discover important resources associated with it.

Aerial and space views of the surface of the Earth frequently can complement each other. The broad synoptic view obtained with space photography can be used to maximum advantage in drawing boundaries that discriminate one type of resource feature from another. Then, through a process known as "multistage sampling," very-large-scale aerial photography can be obtained of small representative areas within each such type in order to identify it.

CHOICE OF PHOTOGRAPHIC FILMS

On the Apollo 9 mission, simultaneous multiband photographs of the surface of the Earth were obtained from space. Every type of feature encountered on the surface of the Earth tends to reflect and emit radiant energy in distinctive amounts at specific wavelengths. Consequently, when remote sensing is done simultaneously in each of several wavelength bands (a process known as "multiband sensing," "multispectral sensing," and "multiband spectral reconnaissance"), each type of feature theoretically becomes identifiable by its multiband "tone signature" or "spectral response pattern."

It was with this possibility in mind that a special photographic team held a series of meetings, extending over more than 2 yr, primarily for the purpose of selecting the three bands that would be most useful in a multiband space photography experiment. Consistent with the recommendations of that team, the three bands used on the Apollo 9 mission in obtaining simultaneous black-and-white photographs from space were those for exposing the green, visible red, and near-infrared wavelengths of radiant energy. The same three wavelength bands are to be employed on ERTS–A, the first in a series of Earth Resources Technology Satellites, which is scheduled to be launched in 1972.

Some investigators, on noting that a color film

known as "Infrared Ektachrome" contains three dyes that, in effect, are responsive to the green, red, and near-infrared wavelength bands, have argued that this single color film might provide all the information obtainable from the corresponding three black-and-white photographs. To evaluate this argument, a fourth camera, containing Infrared Ektachrome film, was used on the Apollo 9 mission simultaneously with the three cameras loaded with black-and-white films. This camera package (consisting of four Hasselblad cameras having 80-mm focal lengths and accommodating 70-mm roll films) was designated by the NASA Photo Team as Scientific Experiment 065 (S065). (See fig. 1–1.)

Other investigators contended that a conventional color film such as Ektachrome would be more useful because it would reveal the surface of the Earth in the normal colors with which men are accustomed to viewing the landscape. This "normality" of appearance, it was realized, would facilitate the recognition of some features that would have an abnormal appearance on either Infrared Ektachrome or multiband black-and-white photography. Hence, several matching frames of photography also were taken with con-

FIGURE 1–1.—Components of the multiband camera system flown on the Apollo 9 mission. In the top center is the camera mount; left to right below it are the four Hasselblad 70-mm cameras, and below them the corresponding film magazines and filters. In addition, a manually controlled switch for activating the shutters simultaneously on all four of these battery-powered cameras appears in this photograph. Film-filter combinations used in these four cameras (from left to right) were as follows: Panatomic-X (3400) with a Wratten 58 (green) filter (Pan-58), Infrared Ektachrome (SO–180) with a Wratten 15 (orange) filter, Panatomic-X (3400) with a Wratten 25A (red) filter (Pan-25A), and Infrared Aerographic (SO–246) with a Wratten 89B (dark red) filter (IR–89B). For a view of the assembly as mounted in the Apollo 9 vehicle, see figure 1–2.

ventional Ektachrome film from hand-held Hasselblad cameras during the Apollo 9 flight.

Both Ektachrome and Infrared Ektachrome films are the "subtractive reversal" type in which the dye responses, when the film is processed, are inversely proportional to the exposures received by the respective layers or wavelengths. Infrared Ektachrome film has three dyes with wavelength responses as indicated in table 1–1. The cyan-coupled dye, which is linked to the infrared-sensitive layer, produces little or no cyan dye on the film when viewed on a light table. Instead, the green and red sensitive layers that are coupled with the yellow and magenta dyes, respectively, come through strongly, and this combination makes anything that has high infrared reflectance and relatively low reflectance of visible wavelengths (e.g., healthy vegetation) appear reddish on the resulting positive transparency.

Alternately stated, the reflectance of healthy vegetation in the green and red portions of the spectrum is sufficiently low to induce a strong response by both the yellow and magenta dyes at the time of processing, but in the infrared part of the spectrum its reflectance is sufficiently high so that little response is induced in the cyan dye at the time of processing. The high concentration of yellow dye in the processed transparency almost completely absorbs blue light; furthermore, the high concentration of magenta dye almost completely absorbs green light. Consequently, when the processed transparency is viewed over white light, healthy vegetation appears red.

Table 1–1 facilitates an understanding both of the means by which a conventional Ektachrome film provides true colors in the processed image and by which an Infrared Ektachrome film provides certain false colors.

An additional factor governing the relative usefulness of Infrared Ektachrome and the corresponding three-band black-and-white photographs is summarized in table 1–2. The information shown there was provided by Dr. P. N. Slater, one of the coinvestigators for the S065 experiment. The table also shows the spatial resolution capabilities of the particular lenses that were used in the four-camera S065 system. In comparing the data of that table with actual S065 photographic results, we conclude the following:

(1) One of the four cameras used in S065 (camera BB), on the basis of the lens system used, was potentially capable of providing the sharpest pictures of all (59.2 line pairs per millimeter when using a high-contrast target, or 280-ft ground resolved distance for low-contrast features on the ground). However, in S065 the lens capabilities of this camera were not fully exploited. Specifically, camera BB was used to photograph green wavelengths of light—the shortest wavelengths used in the experiment. In accordance with the laws governing atmospheric scattering, the shorter the wavelength used in taking aerial or space photographs, the more troublesome the scattering by atmospheric-haze particles becomes and the more blurred the photographic image. Consistent with this explanation, camera BB produced the least sharp images of all four cameras.

(2) The two cameras of the four (AA and CC) that, on the basis of wavelengths exposed for, were capable of providing the sharpest pictures of all, likewise were not able to exploit that capability fully. In these instances, however, it was primarily because the lenses of the two cameras (like those of the other two used in the S065 experiment) had been color corrected so as to take sharp photographs with visible light when focused at infinity. When their focal settings had been increased for the S065 (as necessary to accommodate for infrared wavelengths), their AWAR capabilities had decreased to the values shown in table 1–2. For many terrain features, including healthy vegetation, there is higher infrared reflectance than visible-light reflectance. This greater "scene brightness" is an additional factor favoring higher spatial resolution on infrared photography. However, this factor was offset by the inherently lower resolving powers of the two infrared-sensitive films used in S065.

(3) By operating in a spectral zone where scattering by atmospheric-haze particles is only moderately troublesome, and by using a high-resolution photographic film, camera DD provided the sharpest photographs of all.

In rebuttal to the Infrared Ektachrome enthusiasts, advocates of obtaining three separate black-and-white photographs have cited the following theoretical advantages, all of which can now be evaluated in practical terms because Earth resource test sites were successfully photographed with both

TABLE 1–1.—*Spectral Responses of Normal Color and Infrared-Sensitive Color Films*

Type of film	Spectral region			
	Blue	Green	Red	Infrared
Normal color film (e.g., Ektachrome):				
Normal color film sensitivities	Blue	Green	Red	
Color of dye layers	Yellow	Magenta	Cyan	
Resulting color in photographs	Blue	Green	Red	
Infrared sensitive color film (e.g., Infrared Ektachrome):				
Sensitivities with Wratten 15 yellow filter	(a)	Green	Red	Infrared
Color of dye layers		Yellow	Magenta	Cyan
Resulting color in photographs		Blue	Green	Red

a All 3 layers of Infrared Ektachrome film are sensitive to blue light, but the Wratten 15 yellow filter prohibits blue light from sensitizing any of the 3 layers.

TABLE 1–2.—*Settings and Camera Resolution Characteristics of the S065 System*

Camera designation	Film-filter combination	Focal setting	AWAR [a]	GRD [b]
AA	Infrared Ektachrome, Wratten 15	50	33.4	470
BB	Panatomic-X, Wratten 58	Infinity	59.2	280
CC	Infrared black and white, Wratten 89B	33	32.2	470
DD	Panatomic-X, Wratten 25A	Infinity	43.1	370

a AWAR is the abbreviation for "area weighted average resolution" and in this instance applies to a high-contrast target. It is expressed in this table in line pairs per millimeter for each camera and for the focal setting at which that camera was operated on the Apollo 9 mission. The AWAR values were taken by Keenan and P. N. Slater (1969).

b GRD is the abbreviation for "ground resolved distance" and in this instance applies to a low-contrast target. Figures given are for distance on the ground, in feet, encompassed by one line pair. For linear features of moderately high contrast the effective GRD values may be better than the nominal GRD values listed here by a factor of 10 or more. For example, one-lane roads extending across the desert are frequently discernible on the Apollo 9 S065 photography, even though the width of the entire roadway clearing is no greater than 20 to 30 ft. These GRD values assume a flight altitude of 200 km (103 n.mi.). The GRD values will appear in a final report by Keenan and Slater and are included here by permission of the authors.

the S065 multiband camera system and with handheld cameras loaded with Ektachrome film:

(1) When three black-and-white photographs are obtained simultaneously, but in three separate wavelength bands (as in S065), each film can be given optimum exposure. The importance of this consideration is obvious when we emphasize that the optimum exposure that was required when photographing from space with one of the S065 black-and-white photographs (that exposing for infrared wavelengths) was 1/250 sec at $f/16$, whereas that for another of the black-and-white photographs (that exposing for green wavelengths) was 1/125 sec at $f/4$. This represents an exposure difference in the two bands of five full stops. Consequently, a single color film such as Infrared Ektachrome that seeks to expose for both of these bands simultaneously does not lend itself to any compromise in f-stop and shutter-speed settings that will provide comparably good exposures.

(2) In the black-and-white three-camera system, each lens can be set at optimum focus for the narrow range of wavelengths used when taking pictures with it. Again, actual figures are needed to quantify this potential advantage. In S065, all cameras used had focal-length calibrations emplaced by the manufacturer based on the presumption that normal panchromatic film would be used. Consequently, it was entirely appropriate that a focal setting of "infinity" be used during the Apollo 9 mission on the cameras that were exposing only for green and red wavelengths. However, a setting

of 33 ft (instead of infinity) was found to provide optimum focus for the third Hasselblad camera, because it was exposing only for relatively long wavelengths, which are focused at a much different distance than are the green and red wavelengths. The fourth camera of the S065 system (the one using Infrared Ektachrome film) was focused at 50 ft in an effort to obtain the best compromise in focal settings, because it had to record green, red, and infrared wavelengths simultaneously.

(3) In the black-and-white three-camera system, uniformity of color balance can be achieved. Sequential photography taken with Infrared Ektachrome film may not yield a consistent color balance. Consequently, it is difficult for the photointerpreter to assign a color code to each type of Earth resource feature with confidence that the same hue, value, and chroma characteristics will be exhibited by that feature each time it is photographed with Infrared Ektachrome film. Color combining three separate black-and-white photographs (i.e., three that separately recorded the same three wavelength bands as are used in taking an Infrared Ektachrome photograph) largely overcomes this difficulty because each can be "normalized" separately.

(4) In the black-and-white three-camera system, opportunity is afforded to produce a variety of "false color" images. When Infrared Ektachrome film has been properly exposed and processed, it produces false colors that greatly facilitate the photointerpretation of vegetation. However, if it were possible to assign a different weight and a different color to each of the three wavelength bands to which this film is sensitized, additional false-color enhancements might be made to facilitate the interpretation of certain other Earth resource features. These additional false-color enhancements can, indeed, be made if the same three bands have been recorded separately on multiband black-and-white photographs.

(5) When the black-and-white three-camera system is used repetitively, sequential photographs can be color combined and enhanced. Color Plate 7 of this report illustrates the value of color-combining sequential black-and-white photographs that were taken with the same film-filter combination (in this case, Pan-25A). In other instances, the optimum color coding of Earth re-

source features may best be achieved by color combining a Pan-25A photograph taken on one date with an IR-89B photograph taken on another date. In all such instances the color combining can be readily accomplished if separate black-and-white photographs are available, but the task would be more difficult if only Infrared Ektachrome photography were available. (By three successive filterings of an Infrared Ektachrome transparency, the three dye records it contains can be separated to produce a multiband set consisting of black-and-white records.)

PHOTOGRAPHY OBTAINED FROM APOLLO 9 AND FROM SUPPORTING AIRCRAFT

In S065, a four-camera system was used. (See figs. 1–1 to 1–3.) Each of the 70-mm Hasselblad

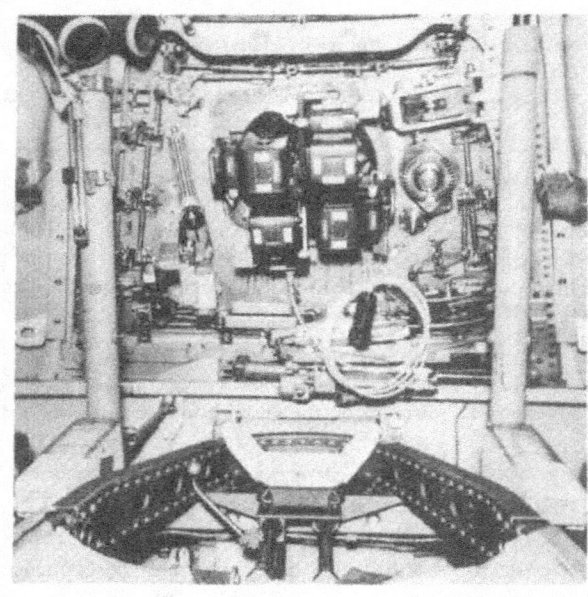

FIGURE 1–2.—In this view the multiband camera system that was flown on the Apollo 9 mission is shown installed over the hatch window of the vehicle, directly forward of the middle couch. Vertical photos were obtained by orienting the entire space vehicle in such a way that the optical axes of all four cameras pointed toward a nadir point on the surface of the Earth. The flight-attitude indicator was set to torque at the spacecraft's orbital rate of about 4 deg/min. The astronaut piloting the spacecraft followed the attitude indicator during a photographic run and in this way the spacecraft's hatch window remained oriented straight down throughout the run.

FIGURE 1–3.—The Apollo 9 vehicle in flight after the command module (top left) had docked with the lunar module. The hatch of the command module is open and the hatch window through which multiband S065 photographs were taken is clearly visible.

cameras comprising the system used a different combination of photographic film and filter. All four cameras had their optical axes alined, and their shutters were triggered simultaneously, so that the same portion of the Earth's surface would appear in all four photographs. The principal areas photographed were Earth resource test sites in the United States.

Many of the test sites also were photographed almost simultaneously from aircraft at altitudes that ranged from a few hundred feet to approximately 70 000 ft. In addition, special sensing devices known as "optical mechanical scanners" were operated from aircraft to permit the nearly simultaneous sensing of several of the test sites in a great many wavelength bands of the electromagnetic spectrum, including several bands in the ultraviolet, visible, near-infrared, and thermal-infrared regions. Finally, in selected portions of these same sites, both terrestrial photographs and on-the-ground measurements of radiant energy emanating from representative Earth resource features also were obtained at the time of the overflights.

The following S065 photography was obtained during the Apollo 9 mission:

Orbit 78: U.S. west coast through Tucson, Ariz., through El Paso, Tex., to Houston-Galveston, Tex., 34 exposures in a strip.

Orbit 92: U.S. west coast and Salton Sea, California, six exposures; Roswell, N. Mex., three exposures; Mississippi River, four exposures.

Orbit 93: Salton Sea, five exposures; Tucson, nine exposures; Matagorda, Tex., three exposures.

Orbit 109: Amazon Delta, Brazil, three exposures. (It appears that the spacecraft may have been too far offshore to obtain more than pictures of the coastal waters.) Toluca, Mexico, volcano, three exposures. Costa Rica, volcano, three exposures. (Here again, the orbital ground track may have been too far from the target. As originally planned, this area was to have been photographed by the hand-held cameras, but the crew volunteered to try photographing the area with the S065 equipment, by proper control of the space capsule's attitude.)

FIGURE 1–4.—This Itek rear-projection viewing equipment was used to examine the 70-mm photography flown on the Apollo 9 mission. The equipment also was used to view the 70-mm and 35-mm sequential photography from each high-altitude mission. Magnifications up to 20 diameters can be profitably exploited in viewing this photography. The large screen immediately to the left of the operator displays one such enlargement. The operator is shown displaying a second photograph of the same area, for comparative purposes, on a smaller viewing screen directly in front of him. It is possible to display both images at the same scale or, as in this instance, at different scales. The automatic focusing capability and automatic film-drive mechanism of this rear-projection equipment greatly facilitates its use.

Orbit 120: Austin, Tex., three exposures; Charleston, S.C., three exposures.

Orbit 121: Colorado River Basin to Tucson, eight exposures in a strip; Anson-Snider, Tex., near Abilene, an agricultural site, three exposures; Appalachian Mountains, three exposures; and South Carolina coast, three exposures.

Orbit 136: San Nicolas Island, California, through San Diego, Calif., to Dallas, Tex., 15 exposures in a strip; Columbus, Ga., out to the Gulf Stream, seven exposures.

Orbit 137: Barbados Oceanographic and Meteorological Experiment (BOMEX) site in the Caribbean, 33 exposures. (Because there were more black-and-white film frames in the experiment, the excess was deliberately programed for the BOMEX site—which is largely water—after the exhaustion of the color and black-and-white infrared films primarily over land sites.)

In addition, a few S065 frames were shot of the horizon during orbital sunrise and sunset.

Three hand-held Hasselblad cameras loaded with conventional Ektachrome film also were used on the Apollo 9 mission to photograph various "targets of opportunity." Several exposures taken with these cameras were of NASA Earth resource test sites and closely matched those taken with the S065 four-camera system.

SEQUENTIAL HIGH-ALTITUDE PHOTOGRAPHY

Long before the launch of Apollo 9, it was recognized that certain Earth resource features can best be inventoried through a comparative analysis of "sequential" space photography (i.e., that taken repetitively of the same area at suitable time intervals). However, there appeared to be little likelihood that such photography could be obtained on future Apollo flights for at least 2 yr. Conse-

quently, NASA arranged for simulated space photography to be flown sequentially at monthly intervals throughout the 1969 growing season, from aircraft operating at very high altitudes (approximately 70 000 and 60 000 ft) using a camera system that would accommodate essentially the same four film-filter combinations as were being used on the Apollo 9 S065.

SPECIAL VIEWING EQUIPMENT

For most of the photographs taken by the Apollo 9 astronauts and on sequential high-altitude aircraft flights, 70-mm film was used. Because of the small scale of the imagery thus obtained, it is very important that proper equipment be available for enlarging and viewing this photography. Fortunately our group was able to obtain on a loan basis and at no cost one of the best viewers available for this purpose, the Itek rear-projection viewer shown in figure 1–4. This equipment permits two 70-mm rolls of photographs to be viewed simultaneously.

Several types of color-combining equipment were used in producing multiband and multidate color composite photographs. Although the color-enhanced images produced by these devices could be viewed directly, color photographs that were taken of them were used in the tests reported upon in chapters 3 and 4. The color-enhancing equipment itself is described in chapter 2.

SELECTED LITERATURE

Colwell, R. N. 1968. The Usefulness of Sequential Space Photography for Earth Resource Inventory. IAF Paper AS163.
Keenan, P. B.; and P. N. Slater. 1969. Preliminary Postflight Calibration Report on Apollo 9 Multiband Photography Experiment S065. Tech. Memo. 1, Optical Sciences Center, Univ. of Arizona.

2

Additive-Color Image Enhancement—Techniques and Equipment

Jerry D. Lent

"Image enhancement" has become a popular term for a number of specialized remote-sensing techniques designed to help analysts extract useful information from multiband images. This section will describe some of these techniques and the equipment used when implementing them. The potential usefulness of such techniques for the inventory of Earth resources will be indicated only in a general way here, but will be illustrated specifically in other chapters.

The purpose of any form of image enhancement usually is twofold: (1) to increase the total amount of information that is derivable from the "raw data" and (2) to facilitate the data extraction process, diagramed as follows:

RAW DATA
(Photographs or other images from remote sensors)

Conventional photo-interpretation and analyses → Enhancements of the raw data

DERIVABLE INFORMATION

DERIVABLE INFORMATION

When additive-color image-enhancement techniques are employed, these objectives usually are realized by highlighting subtleties in image tones. Such enhancements can be accomplished by either optical or electronic means.

BASIC PRINCIPLES OF THE ADDITIVE-COLOR PROCESS

The additive-color technique for multiple-image enhancement has been known for many years. Only recently, however, have certain specific techniques and equipment been developed that show promise enough to attract the attention of scientists from a wide variety of disciplines. The potential usefulness of some of these new additive-color-process adaptations is best appreciated if we first examine an elementary principle that is illustrated in Color Plate 1(a). In producing that illustration, three colored light beams, one red, one blue, and one green, were projected onto a screen in such a way that only a small portion of each was allowed to overlap the other. This particular additive-color-combination experiment reveals that "new" colors can be produced in such a manner. On the composite image, sharp color differences and clearly defined boundaries serve to differentiate three types of areas: (1) those in which only the original color beam is present, (2) those in which an additive-color combination is produced as a result of overlapping pairs of beams, and (3) a region in which the additive-color combinations of all three primary colored beams results in white light. Many other color combinations are possible: four-color and even five-color combinations are also feasible and might prove useful for certain experiments.

The most common form of additive-color enhancement employs an optical "combiner" and involves the use of a set of black-and-white multiband images (transparencies) all showing the same scene. Each image is projected by means of a light beam that has a particular hue because a colored filter is placed in its path. These color images are then superimposed on the viewing screen to form a single color image. The colors thus exhibited on the composite image are a function of (1) the image densities on the respective multiband transparencies and (2) the color values of the filters through which the transparencies are projected.

LIMITATIONS AND ADVANTAGES

There are limitations as well as advantages associated with the use of additive-color enhancement by those who seek to inventory Earth resources from multiband photography. The main limitations are in (1) the extent to which certain enhancements can be accomplished, (2) need for special equipment with which to perform the enhancements, and (3) requirements for special interpreter training.

The image analyst should not expect miracles from additive-color enhancement. Enhancements are usually performed to highlight *differences* between multiband images. The mere addition of various colors (when projecting and superimposing images that were obtained under identical conditions of time, exposure, and film-filter selection) does not invariably increase data extraction capability. If there are no density differences between the multiple images that are to be studied, there is not much point in attempting image enhancement. At the other extreme, if density differences between the multiple images appear to be very large, there may be no need for enhancement because of the ease with which such differences can be discerned without it. Enhancement usually can prove most useful when the data sources have relatively subtle density differences that may go undetected under visual gray-level discrimination. Optical and electronic enhancements may reveal these subtle density differences.

To use color-enhancement techniques successfully, the image analyst must become familiar with special equipment. Some of this equipment is costly. Most of the current elaborate color-enhancement systems, moreover, are prototypes of systems that will undoubtedly undergo refinements requiring further familiarization before they can be readily used by a skilled image analyst.

Finally, the image analyst must be trained in the skills of interpreting additive-color-enhanced images that often exhibit "unconventional" distortion, resolution, and color display characteristics.

Some scenes may be very confusing to the interpreter at first, despite their potentially valuable data content. Numerous studies have been conducted on human color visual responses, but relatively little has been done to determine the most suitable color contrasts for discriminating specific features as imaged on multiband photography.

The most obvious advantage to be gained from the use of a color-enhancement technique is the ability to combine the information (either electronically or optically) from several images into one single frame for interpretation purposes. Mentally integrating these multiple images may be a difficult, if not impossible, task for the human mind unless the images are themselves integrated and color coded as in the additive-color process. As previously indicated, this technique facilitates the recognition not only of *spectral* differences on multiband photography taken on a single date, but also of *sequential* differences on multidate photography taken in a single band. It follows that various combinations of spectral and sequential differences also may be profitably enhanced in certain instances. As indicated elsewhere in this report, any differences in the data discernible on sequential photography ordinarily can be attributed to actual changes in a feature's condition or state. However, this assumes, of course, that the remaining factors affecting the recorded tone signatures (exposure, processing, etc.) are sufficiently standardized. Some examples of "sequentially enhanced" color composites appear in chapter 3.

Electronically controlled additive-color enhancement systems, although considerably more complex and expensive to develop than optical systems, offer the image analyst greater flexibility in interpreting his data. Very subtle gray-level differences can be "sliced" and electronically "expanded"; then digital codes can be ascribed to them for subsequent analysis. Also by electronic means, different color codes can be assigned to specific input densities, and color enhancement can be performed by this means. Similarly, the information recorded on the individual layers of the color-film emulsion can be directly coded and redisplayed with different colors if necessary for enhanced interpretability. Consequently, greater flexibility of color display may be possible with electronic than with optical enhancement systems. The best features of both the optical and electronic systems can be incorporated in one piece of equipment by still further development, as will presently be described.

IMAGE-ENHANCEMENT SYSTEMS EMPLOYING COLOR IN THEIR DISPLAYS

The systems to be described first are those with which the author has had considerable experience. These systems are believed to represent the current state of the art.

Forestry Remote Sensing Laboratory Optical Color Combiner

An unsophisticated but effective color combiner used in one form or another by several groups is exemplified by one at the NASA–USDA Forestry Remote Sensing Laboratory (FRSL), University of California, Berkeley. Figure 2–1 diagrams the technique employed with that equipment in optically combining multiband images and projecting them through color filters to produce color-enhanced scenes.

The advantages of this optical system are (1) its ease of construction and use (the equipment is inexpensive and readily obtainable, and the enhancements are readily performed); (2) the data

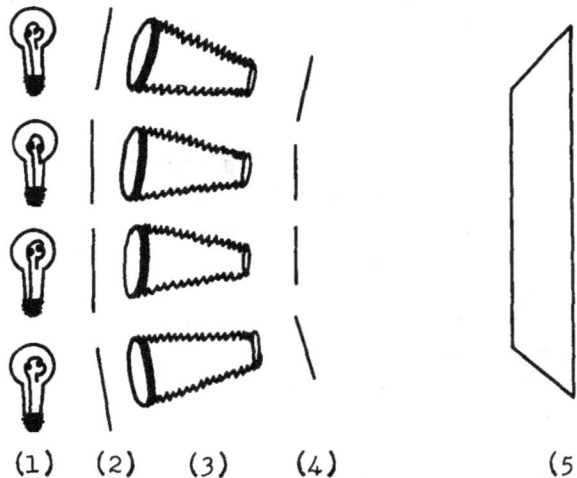

FIGURE 2–1.—Optical color combiner. The light source is shown at (1), the multiband images at (2), the optical train for projection at (3), and the color filters at (4) through which the images are projected onto a front-viewing or rear-viewing screen at (5).

can be directly interpreted from the screen; and (3) there is a large selection of available colored filters, and thus a large number of color combinations with which to experiment.

Limitations are mainly related to the deficiencies encountered in attempting to use stronger light sources for more flexibility of color combinations (since some of the color filters have reduced transmission characteristics when used with low-level illuminants). There is also the problem of obtaining adequate registration of images for color enhancement; slight differences in geometry between images can cause density combinations that are confusing to interpret and may lead to errors. In addition, a problem can result from the use of uncalibrated data of variable quality. The image densities to be enhanced must have some determinable, quantitative relationship to the features they represent; hence, changes in time of day or date of photography, Sun angle, film type, or even processing procedure can cause shifts that alter this relationship.

Color Plate 1(b) and (c) show examples of the output from this color combiner. The photograph in (b) showing a portion of the Imperial Valley of California was produced by registering three black-and-white transparencies obtained on three separate dates: March 8, April 23, and May 21, 1969. The various color codes are attributed to changes in tone signature on the transparency between those dates. Fields in black denote continuously vegetated fields on all three dates; red to orange denote row crops that were harvested by the last date; green to blue denote mowed cover crops; and yellow to chartreuse denote predominantly bare regions on all three dates.

Color Plate 1(c), showing 16 sq mi of agricultural cropland near Phoenix, Ariz., is an example of the FRSL optical system, the system having the best resolution.

Two other systems (described below) provided the other examples. The Philco-Ford viewer (Color Plate 1(e)) has a finer resolving scanner and yielded the next best resolution. The image discrimination, enhancement, combination, and sampling (IDECS) example (Color Plate 1(d)) has the worst resolution, but that system is often of great value because of the large number of multiband image or sequential images that can be combined.

University of Kansas Electronic Processing System

An electronic image-enhancement and correlation system designed and built by personnel at the CRES Laboratories, University of Kansas, is quite different from the optical system described above. This electronic system, IDECS, was designed as a flexible research tool. Figure 2–2 shows the console and viewer.

With the IDECS device, image enhancements can be accomplished either among monochromatic combinations of multiple images or by additive-color combinations of them, and the operator may "level select" for tone or brightness range on one or more images for purposes of highlighting encoded spectral densities. Isodensity enhancement can be performed and, ultimately, directly analyzed by online computer software. Input images to be interpreted can be from different imaging sensors, of somewhat different geometry (because "reasonable" linear distortions between images can be corrected electronically), and can be as large as 3 by 4 in. The output display recording is best performed directly from the viewing screens, but adequate documentation is often possible by photographing the viewing screens with certain color films.

Some of the features of IDECS provide interpretation flexibility that optical systems cannot duplicate; on the other hand, for problems where spatial resolution may be very important, an optical system usually is superior to the IDECS system. Where large amounts of multiband imagery must be analyzed quickly, the IDECS system is far more efficient than current optical systems. With appropriate hardware implementation, far more images can be effectively combined by this system than by optical systems. The IDECS system is currently designed to accommodate six images, but this is not a limit to the number of images that eventually might be simultaneously analyzed by such a system. An example of output from the IDECS console viewer appears in Color Plate 1(d).

Philco-Ford Image-Tone Enhancement System

A system that incorporates both optical and electronic capabilities for enhancement purposes is the Philco-Ford console viewer shown in figure 2–3. With it, two multiband images can be illuminated by a single source of light for projection and scanning. Beam-splitting optics provide this feature, which eliminates any concern for variability in relative intensities of two or more illumination sources. This feature also reduces the number of flying spot scanners that the system requires. The scanner employed in the Philco-Ford viewer can resolve a total of approximately 1000 lines, whereas the IDECS system currently in use employs scanners that can resolve only about 500 lines (the same number commercial television displays).

FIGURE 2–2.—IDECS console and viewer arrangement. Dalke and Estes (1968) have prepared a complete technical and operational description of this system.

FIGURE 2–3.—Philco-Ford two-channel color-enhancement console viewer. The viewing aperture (approximately 5 by 7 in.) is in the dark area to the left of the console controls. These controls operate the mixing and digitizing of the two channels of imagery.

A special color wheel used in this system results in the additive-color mixing of red and green hues. Sixteen levels of density can be referenced within each channel and "digitized" for subsequent quantification. The "level select" feature is nominally linear and can be expanded and contracted as with the IDECS system; each density level can also be modified should the image analyst prefer to weigh its importance in the display. An example of the output from this console viewer appears in Color Plate 1(e).

Since the Philco-Ford system presently accommodates only two bands of photography at a time, mention should be made of a means by which its capability could be improved to accommodate, in effect, four bands. Let us presume that as many as six or eight bands of black-and-white photography have been obtained simultaneously of an area in which we wish to accomplish multiband image enhancements with the Philco-Ford device. From a study of the tonal characteristics on each of these simultaneous exposures, we are able to select the four bands on each of which unique tone values appear for certain Earth resource features. For image-enhancement purposes all of the remaining negatives are discarded because, on those discarded, the tone values for objects that are to be identified differ the least from tone values of other negatives in the series. The remaining four are paired off to constitute two pairs of negatives.

Obviously, if the two negatives of a pair exhibited exactly the same tone characteristics, there would be no gain from using the two of them instead of merely one. Since it is therefore the *differences* in tone between the two members of a pair that are most likely to lead to the correct identification of objects, the negative transparency of one member of each pair is superimposed on a positive transparency that has been made from the other member of the pair. When the two transparencies are in proper register and light is directed through them, the bright images of one transparency are exactly offset by the dark images of the other transparency, except for objects having differences in reflectance in the two spectral bands that were used in obtaining the two spectrozonal photographs.

Exploiting only these differences, a composite black-and-white negative is made from each of the two matched pairs. The two resulting composite

FIGURE 2–4.—The viewer portion of the Long Island University multispectral enhancement system. Controls for increased color saturation and brightness as well as color selection are readily accessible. The viewing screen is approximately 4 by 9 in. The complete system is designed to operate most effectively with a special four-lens camera, not shown here. The optics of the camera are matched to the optics of the viewer such that a 9-in. reel of aerial film can be accurately registered in a minimum of time.

negatives thus highlight differences in light reflectance in a total of four spectral zones. These two composite negatives (or positive transparencies made from them) could then be used in the Philco-Ford device. While a similar process could be used in either the FRSL optical combiner or the IDECS system, the need is not so great because they can accept a larger number of original multiband transparencies as direct inputs.

Other Systems

A few other systems warrant mention here because of their unique capabilities. The Long Island University multispectral camera-viewer console is shown in figure 2–4. This system is designed to minimize registration problems from which most optical systems suffer. Still another type of multiband image-enhancement system is exemplified by equipment designed by Technical Operations, Inc. (Tech/Ops), of Burlington, Mass. This optical enhancement system enables color-image retrieval from a single emulsion plane. Through a novel "camera" design that incorporates specially ground gratings (spatial carriers) of extremely fine resolution, a number of multiband exposures can be placed on the same emulsion plane and later retrieved and colors assigned for display purposes. The console viewer for this system is illustrated in figure 2–5. The problem of registration is eliminated with such a device. The camera can be used either to record single images of many different scenes or, for multiband work, it can be filtered with the gratings in such a way as to record multiple images of a single scene, each in a different wavelength band.

Scientists in many laboratories are experimenting with devices and techniques that closely resemble those described here. RCA, General Electric, and Hycon, for instance, have developed very sophisticated color-enhancement systems. A few "hybrid" optical systems similar to the FRSL apparatus also can be found in various research laboratories. Some employ a 3¼- by 4-in. image format, whereas others employ 35-mm transparencies. At present all of these devices or systems can be regarded basically as research tools. Electronic systems that automate the enhancement procedure may eventually give interpreters online computing capability.

PRACTICAL APPLICATIONS

At present, research is underway to refine and standardize additive-color enhancement techniques to the extent that predictable, consistent results can be obtained from future experiments designed to inventory Earth resources. Once the technical problems have been solved with respect to (1) which wavelength bands are most useful for each particular terrain feature of interest (and surely there will be different recommendations for the various resource types), (2) which color combinations and enhancements yield maximum information for the objectives of the experiment, and (3) what the gain in benefits over conventional interpretation techniques are for certain applications, then practical, operational use of data-reduction and data-enhancement techniques will become routine.

Efforts to solve problems in the above three areas continue to be made at a number of institutions, including the FRSL at Berkeley. Processors of

FIGURE 2–5.—A relatively portable console viewer or "color reconstitutor" for use with the Tech/Ops camera output. The multiexposed emulsion is placed in a holder (lower left side) and viewed through the monocular. Color input controls appear on the panel at the right. Other color controls are possible with slight modification to the viewing unit for more false-color manipulation. Tech/Ops is also considering producing units that have more than three input color controls.

remote-sensing data acquired for agricultural areas should be able to identify crops and estimate yields by enhanced tones or hues. Such capabilities would materially assist in the continued development of a strong agriculture program policy. Enhanced imagery can assist foresters and other wildland managers in evaluating complex multiple-use problems. Hopefully, enhanced imagery soon can be used operationally in timber typing, detecting stand infestations, locating recreation sites (or more importantly, sites *not* suitable for recreation or specific purposes), in delineating soil boundaries, and in making management decisions such as brush conversion and planting.

SELECTED LITERATURE

DALKE, G. W.; AND J. E. ESTES. 1968. Multi-Image Correlation Systems Study for MGI. Final Rept. Contract no. DAAK02-67-C-0435. U.S. Army Engineering Topographic Laboratories, Fort Belvoir, Va.

Analysis of Earth Resources in the Phoenix, Arizona, Area

David M. Carneggie, Lawrence R. Pettinger, and Claire M. Hay

On March 12, 1969, Apollo 9 obtained Infrared Ektachrome and black-and-white multiband space photographs over Phoenix, Ariz., from an altitude of 126 n. mi. (145 stat. mi.) at approximately 8:30 a.m. local time. On the same day a high-performance aircraft obtained aerial photographs, using essentially the same film-filter combinations, from an altitude of approximately 70 000 ft above sea level.

High-altitude aerial photographs also were obtained at approximately monthly intervals between March and December 1969, for the purpose of studying such time-variant phenomena as agricultural crop development and transient conditions of range vegetation in the Phoenix area. The resultant space and sequentially obtained aerial photographs procured in this photographic experiment have provided a unique opportunity to examine the usefulness of such photography for Earth resource inventory and land-use planning.

The area studied includes a variety of wildland and cultural features and is imaged on three adjacent Apollo 9 photos (AS9–3800, –3801, and –3802). The area extends approximately 100 mi both east and west of Phoenix. (Phoenix appears near the middle of Color Plate 2.) The Forestry Remote Sensing Laboratory (FRSL) has determined the interpretability of space and sequentially obtained aerial photographs of this large area. A major objective has been to determine specific applications and/or limitations of such photography for evaluating the resources of (1) agricultural lands, (2) rangelands, (3) geologic and hydrologic units, and (4) urban and suburban areas.

Within this large study area, small representative test sites were selected where detailed information was collected regarding the condition and identity of features related to each of the land-use categories. Ground data were acquired both at the time of the Apollo 9 overflight and on subsequent dates coinciding with the high-altitude flights. This ground information, when coupled with the analysis of the sequentially obtained aerial photos, has provided the basis for evaluating the merits of the Apollo 9 space photos, and determining their usefulness, either alone or in concert with accompanying high-altitude aerial photos, for making Earth resource analyses in the Phoenix area.

In this chapter the procedures used in attempting to recognize various land-use categories (agricultural, range, geologic, hydrologic, and cultural) will be discussed separately, because different interpretation techniques were employed to study the space and aerial photos depending upon the nature or characteristics of the particular resources.

In evaluating the agricultural resources, the following procedures were used:

(1) Sequentially obtained high-altitude aerial photos were examined individually and in concert for changes in crop development that might produce unique signatures that would facilitate crop identification.

(2) On the basis of information obtained from a study of these photos and from the concurrent acquisition of field data, a crop calendar was prepared.

(3) Tonal variations of the crop types were evaluated from light-transmission data obtained using a Welch Densichron.

(4) Color composites were made from multiband images and from multidate images, in an effort to evaluate the extent to which such composites might improve interpretability of crop types.

(5) Interpretation tests were prepared and given to 12 interpreters, each taking four or five of the 15 different tests. The test images included Infrared Ektachrome and black-and-white photos from both the spacecraft and high-altitude aircraft, sequen-

tially obtained Infrared Ektachrome aerial photos, and color composites made from the black-and-white multiband space and aerial images.

In evaluating the range resources, a two-step procedure was employed:

(1) Representative test sites were selected wherein the species composition, density, and relative palatability of the native forage plants were documented.

(2) The close correlation known to exist between landform, soil type, water availability, natural vegetation type, and plant development was exploited to demonstrate the use of the concept of "convergence of evidence" as an aid in analyzing the forage potential on these sites from space and high-altitude aerial photographs.

In evaluating geologic and hydrologic resources, a three-step procedure was followed:

(1) Representative test sites were selected in which the photogeologist, using available ground data, became familiar with the image characteristics of geologic and hydrologic features seen on the S065 space photography.

(2) Geologic and hydrologic mapping was then performed for other portions of the study area on each of the film-filter combinations.

(3) The results thus obtained were field checked to determine the extent to which geologic and hydrologic mapping could be performed, and on which film types it could be most easily and accurately accomplished.

In evaluating land-use patterns and cultural features:

(1) Broad land-use categories were delineated on a space photo (Color Plate 2).

(2) Cultural features were identified and their relationships to wildland resources were discussed.

THE AGRICULTURAL RESOURCE

The extensive agricultural area under cultivation in and around Phoenix is conspicuous in Color Plate 2. Most of this agricultural land is in Maricopa County, the largest agricultural producing county in Arizona. It contains about one-fourth of the State's total agricultural land. The annual income from agricultural production in this area exceeds $200 million, making agriculture Maricopa County's most important and profitable resource. To evaluate such a resource, crop-area measurements must be made, crop types identified, and

determinations made of crop vigor and yield. This information is vital in preparing an agricultural census, and in the administration of Federal and State agricultural programs.

From our analysis of the S065 space photographs, it appears that acceptable estimates of cropland area can be made by either of two methods. First, the total area designated as agricultural land can be delineated on rectified space photos, and an area-calculation device can be employed to determine the total gross acreage within the delineated area. Allowance can then be made for the proportion of agricultural land not utilized in crop production; e.g., for roads, irrigation facilities, farm buildings, etc. Second, land area under cultivation can be estimated by making a complete tally of the number of individual fields that can be discerned on the space photo.

Identification of crop types from space and aerial photographs is a more difficult task, because of variations in the appearance of crops resulting mainly from different planting times and environmental factors. An effort was made in this study to determine the extent to which crop types and conditions can be interpreted on various kinds of space and aerial photographs. Such a test was made possible by virtue of having obtained the complete field-by-field identity of agricultural crops within a 16-sq-mi test site located just south of the town of Mesa, Ariz. (See fig. 3–1 and Color Plate 3.) This site was selected because (1) it was contiguous, (2) it was easy to reach for gathering crop data on a field-by-field basis, (3) it contained all the important field crop types in the Phoenix area, and (4) it was imaged clearly on Apollo 9 and most subsequent high-altitude photo missions. Crop types and crop conditions (whether seedlings were emerging, crop was mature, or crop was harvested, etc.) were mapped, field by field, within this large test site at the time of the Apollo 9 overflight and at later dates corresponding to the dates of the sequentially obtained aerial photos. This information provided the basis for making the crop-type maps seen in figures 3–2 to 3–16. The crop data also provided the information needed for analyzing the sequential photographs and for preparing interpretation tests designed to determine how well interpreters could identify crop types on different types of space and sequentially obtained aerial photographs.

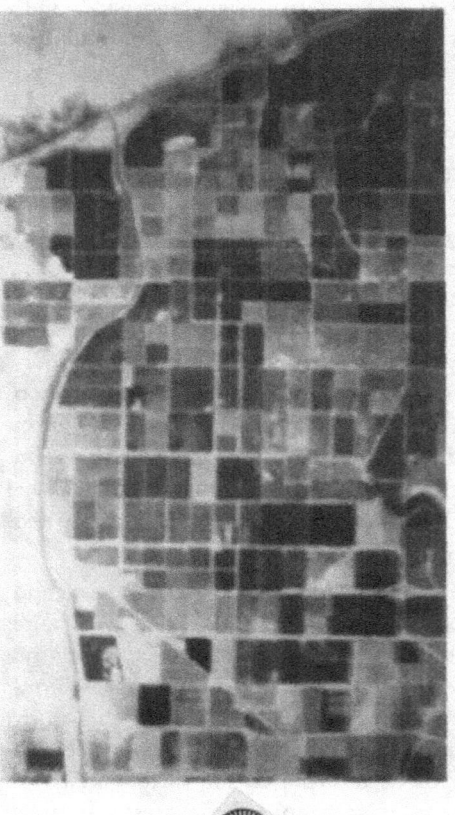

(b) This optically produced additive-color enhancement indicates the potential application of "sequential" tone signature analysis.

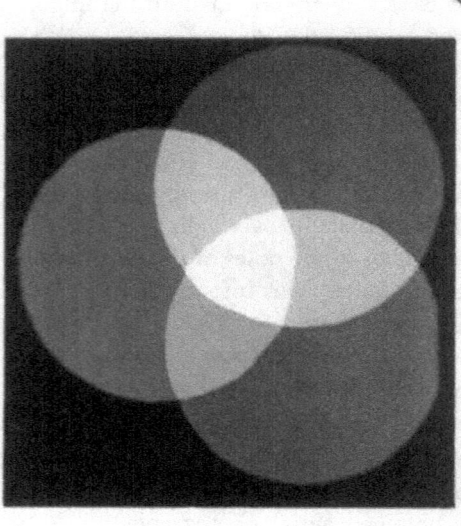

(a) A simple demonstration of additive-color combination. Where all three light beams overlap, white light is produced.

(e) Color enhancement by the Philco-Ford image-tone enhancement device.

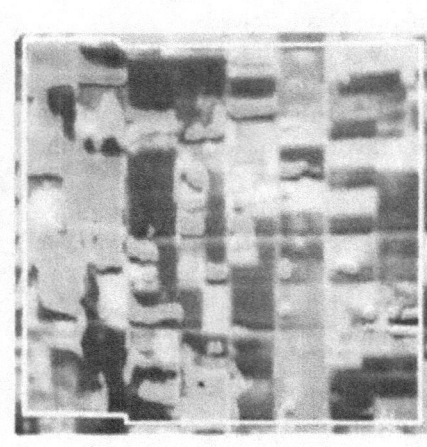

(d) Color enhancement by the IDECS electronic color display and analysis system.

(c) Color enhancement by the FRSL optical color combiner.

Color Plate 1

Color Plate 2. Infrared Ektachrome photograph (enlargement of frame AS9–3801) taken by the
Apollo 9 astronauts. The enlarged space photo and overlay show natural cultural resources in the
Phoenix, Ariz., study area. Within this area (approximately 7000 sq mi) representative test sites
were established to gather detailed information regarding the identity and condition of the
various resources. The area occupied by various land-use categories was determined by an area
calculator as follows: agricultural cropland (A), 20 percent; rangeland (R), 43 percent; uplands
and mountains (M), 24 percent; watercourses (W), 8 percent; and urban (U), 5 percent. (The
actual estimate of cropland realistically needs to be scaled down by a factor of about 10 percent
to account for the roads and farm buildings that are included within the cropland area.)

Infrared Ektachrome space photo (*above*) and high-altitude aerial photo (*below*) taken on March 12, 1969.

Pan-25A space photo (*above*); IR–89B space photo (*below*).

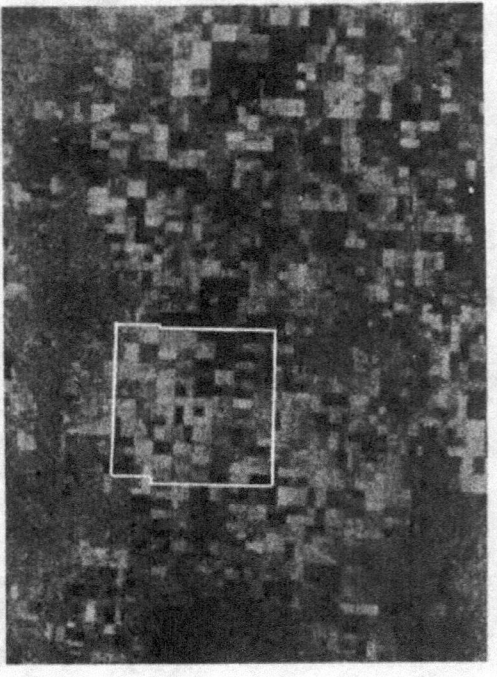

Color Plate 3. These were among photos examined by image analysts to determine how accurately crop types and conditions could be identified within the agricultural test site near Mesa, Ariz. (outlined on photos). The Infrared Ektachrome space photo was test image 6 in the interpretation test described in the text. The Infrared Ektachrome aerial photo was test image 7. The Pan-25A and IR–89B space photos were test images 1 and 2, respectively. The results are summarized in table 3–16.

Color Plate 4. Representative ground photographs showing fields on three dates when high-altitude photographs were obtained. (See Color Plate 5 for field locations.) These photographs aid in the interpretation of high-altitude photographs in Color Plate 5. For example, on the April 23 Infrared Ektachrome image (Color Plate 5) the cotton field appears bare, the barley field has lost some infrared reflectance, and the alfalfa field temporarily has lost almost all its infrared reflectance. On the May 21 Infrared Ektachrome photograph the barley field has a unique whitish color, the cotton field still appears as bare soil, and the alfalfa field again looks dark.

May 21, 1969

April 23, 1969

March 12, 1969

November 4, 1969

September 30, 1969

August 5, 1969

Legend

a	alfalfa
b	barley – bare soil present in November
bs	bare soil
c	cotton
sb	sugar beets – bare soil present after harvest

Color Plate 5. These Infrared Ektachrome photos show the 4- by 4-mi Mesa test site on the dates indicated. All prints are copies from the 35-mm Nikon images except the one for March 12, when the 70-mm-format Vinten cameras were used (note higher resolution of March 12 image). Representative fields of three important crop types and bare soil are indicated on the March 12 photo.

(a) These two color composite images were made using the FRSL optical combiner for the purpose of enhancing overall interpretability of the images. The image on the left was made using IR–89B and Pan-25A Apollo 9 images projected through red (25) and green (61) filters, respectively. The image on the right was made by projecting IR–89B, Pan-25A, and Pan-58 high-altitude aircraft images through purple (35), dark green (74), and dark red (70) filters. These color composites show considerably greater detail than the color composites made by other enhancement systems. Results from interpretation of these images are summarized in table 3–16.

(b) These color composite images were made by the Philco-Ford enhancement device. The composite on the right was made using Pan-25A and Pan-58 high-altitude aircraft images and the left image was made using Pan-25A and IR–89B Apollo 9 images. The differences between the two images are largely due to the resolution.

Color Plate 6

Barley

March 12; 90; yellow
May 21; 75; blue

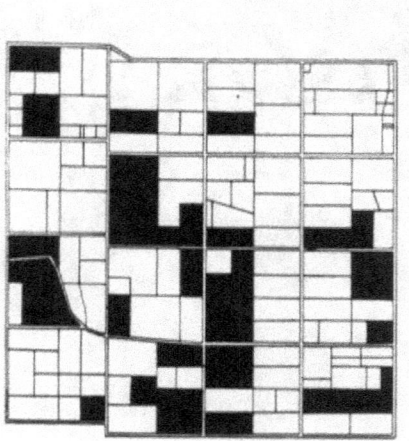

Cotton

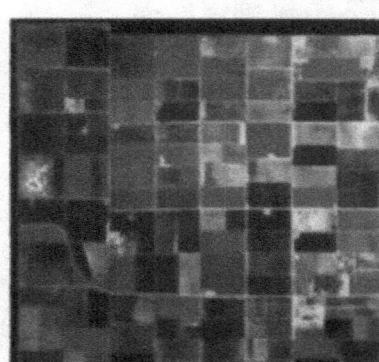

March 12; 75; blue
September 30; 25; red

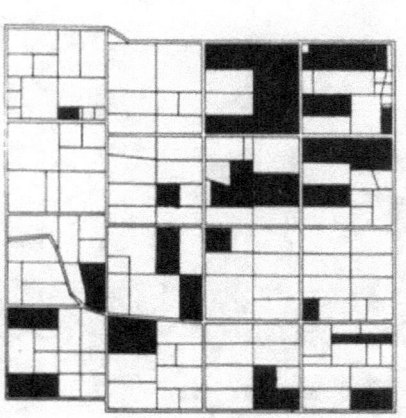

Alfalfa and sugar beets

September 30; 90; yellow
May 21; 35; purple

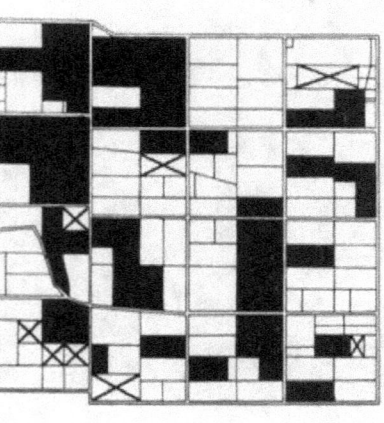

■ Alfalfa
⊠ Sugar beets

Color Plate 7. Color composite images of the Mesa, Ariz., test site prepared with the FRSL optical color combiner. The notations below each image indicate the dates of Pan-25A images that were used, the Wratten filter through which each image was projected, and the color of the filter. The map below each image indicates the location of each field in the category for which the composite image was prepared. The degree of success with which a category can be identified is indicated in table 3–4.

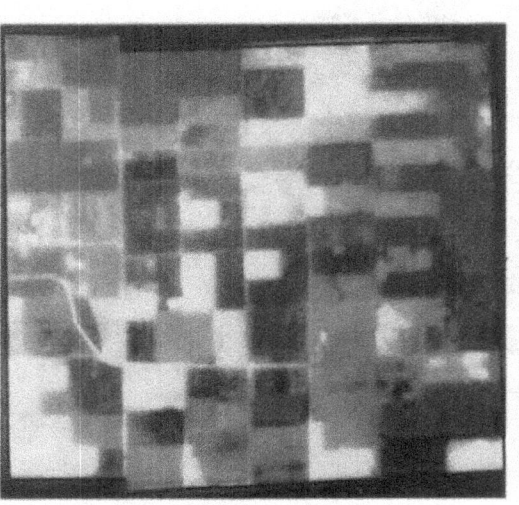

(a) March (Pan-25A) + April (Pan-25A)

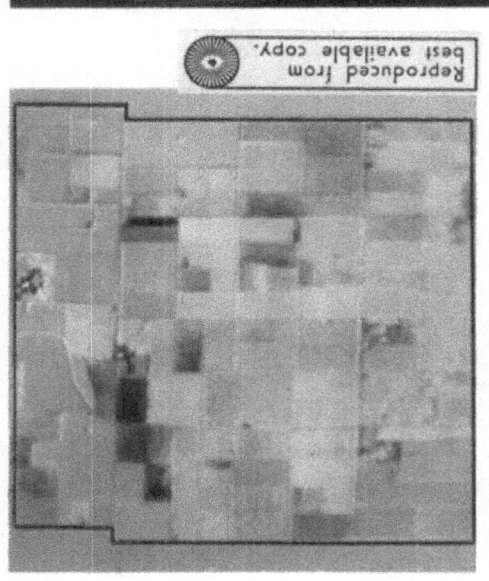

(b) March (IR-89B) + April (IR-89B)

(c) March (Pan-25A) + May (IR-89B)

Three examples of multidate color composite images produced by the Philco-Ford system. The meanings of the color signatures of the various images are tabulated in table 3-5. Pan-25A images were used to produce image (a), IR-89B images for image (b), and one image from each of those bands for image (c).

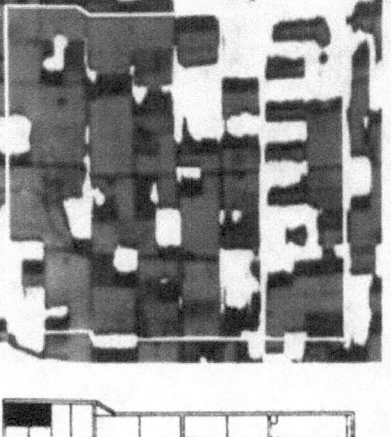

(d) March 12 (Pan-25A + Pan-58 + IR-89B)

(e) March 12 (Pan-25A + Pan-58 + IR-89B)

Color composite images of the 16-sq mi Mesa, Ariz., test site prepared at the University of Kansas using the IDECS system. These images are multiband enhancements made from Pan-25A, Pan-58, and IR-89B images taken on March 12 only. Image (d) was prepared to separate all barley fields from other crops as a unique blue color, and image (e) to differentiate bare soil (pale blue color) from crops. The maps to the right of each photograph show where the fields were in the test site.

Color Plate 8

(a)   (b)

Color Plate 9. Infrared Ektachrome photos copied from a portion of two frames of HyAc
panoramic photography taken on March 9, 1969, from an altitude of approximately 70 000 ft.
Photo (a) shows the influence of low soil fertility on citrus orchards in an area northeast of
Mesa, Ariz. The arrow indicates where gravelly soils from old stream channels do not support
vigorous citrus trees. Photo (b), which covers an area 20 mi west of Phoenix, illustrates further
how different soil types and soil condition can be detected on high-altitude photographs. Three
soil types are delineated on this photo: (1) old alluvial soils, (2) recent alluvial soils, and
(3) gravelly river-channel material. Fields that are still affected by alkali accumulations and
have not been reclaimed for cultivation are indicated at A.

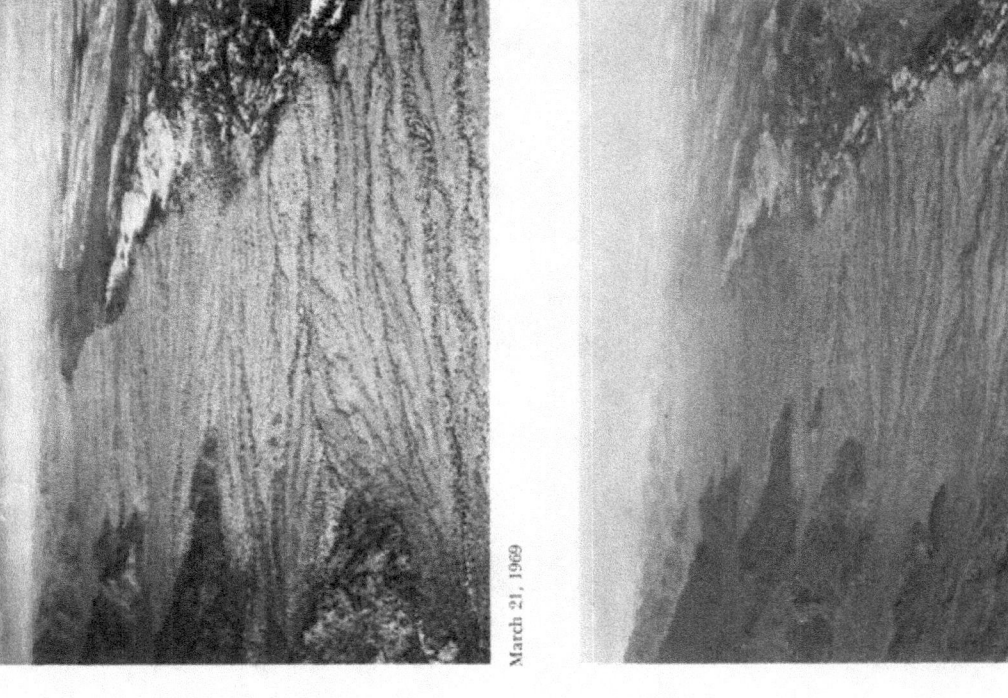

March 21, 1969

May 23, 1969

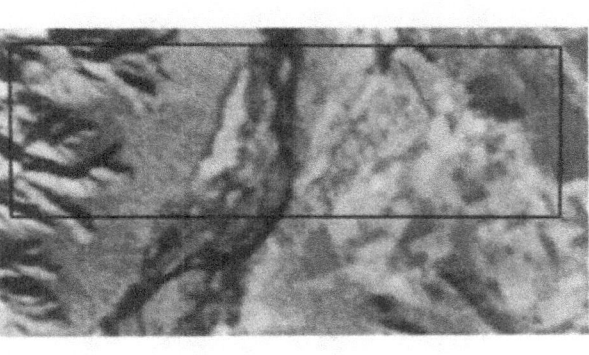

March 12, 1969
Apollo 9

March 12, 1969
high altitude

Color Plate 10. The Infrared Ektachrome Apollo 9 photo enlarged 14 × (*above center*), shows an area of cropland and wildland south of Phoenix. The area outlined was imaged on the same day as the Apollo 9 overflight using Infrared Ektachrome film in the panoramic camera (*above left*). The higher resolution of this image is a decided aid for interpreting the space photo. Certain boundaries, however, are more apparent on the space photo (*i.e.*, the stream-bottom vegetation). The Infrared Ektachrome aerial oblique photo (*above right*) shows a view of a portion of the area outlined on the space photo and was taken shortly after the Apollo 9 experiment to document the stage of development of the vegetation. The oblique view (*below right*) shows the same area on May 23, 1969. In March mesquite tamarisk vegetation had not begun to leaf out, but by May 23 full leafing had occurred. More examples from this area can be found in Color Plate 12 and figure 3–23.

March 12, 1969

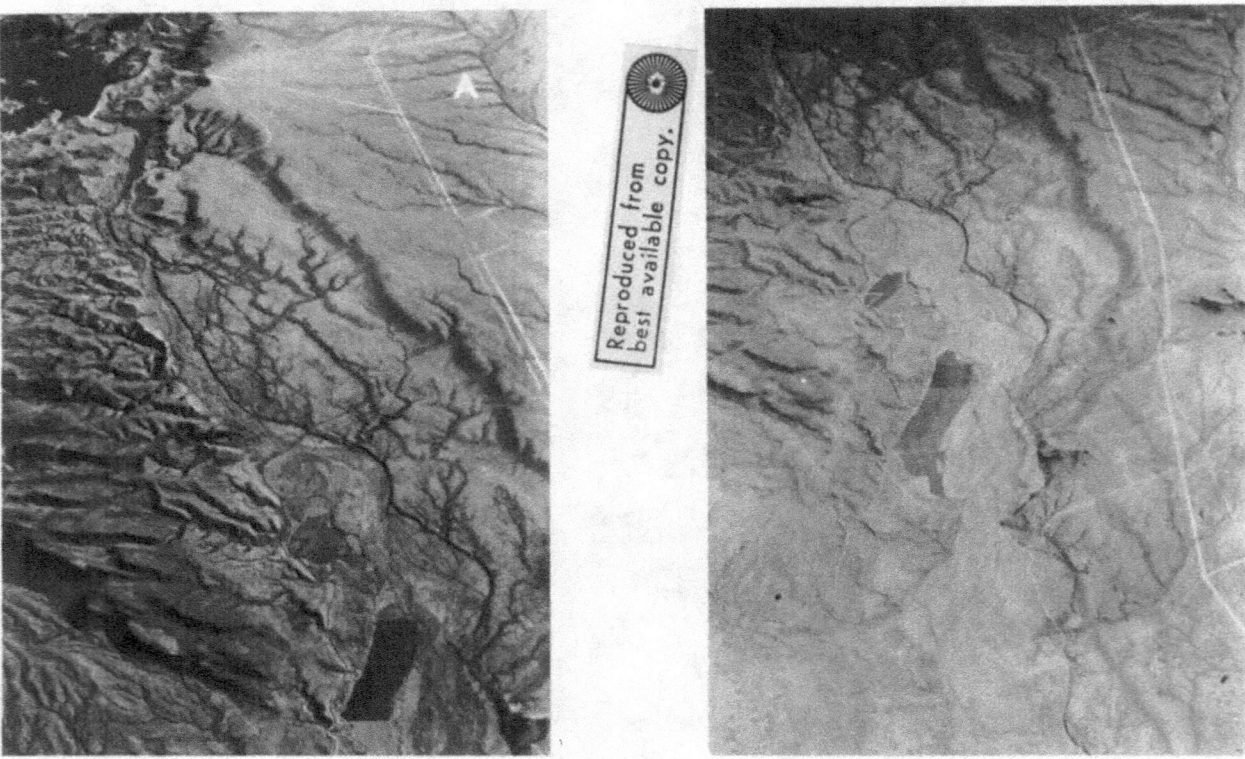

March 12, 1969

April 23, 1969

Color Plate 11. The wildland area in the Infrared Ekta-chrome photo *(top)* is just north of Phoenix. This enlargement was made from an Apollo 9 space photo (AS9–3801), taken from an altitude of 126 n. mi. The rectangle indicates an area that corresponds to the aerial photographs (scale 1/60 000) on this page. Primary use of this land is for grazing, but there is limited recreation use in and around the reservoir. The rugged upland areas (upper portion of space photo) support chaparral vegetation, while the alluvial plains (middle portion of space photo) support semidesert shrub vegetation. Variation of the red coloration associated with the alluvial plains area is due in part to vegetation types associated with different soils, but the more intense red is due to a dense ground cover of annual grass that was at the peak of its foliage development. The area indicated by *A* in the March 12 aerial photograph *(bottom left)* has less vegetation than the other areas. Interpretations such as this may indicate the time of season that a range is ready to be grazed (range readiness), when the forage has reached its peak, and when it has begun to dry.

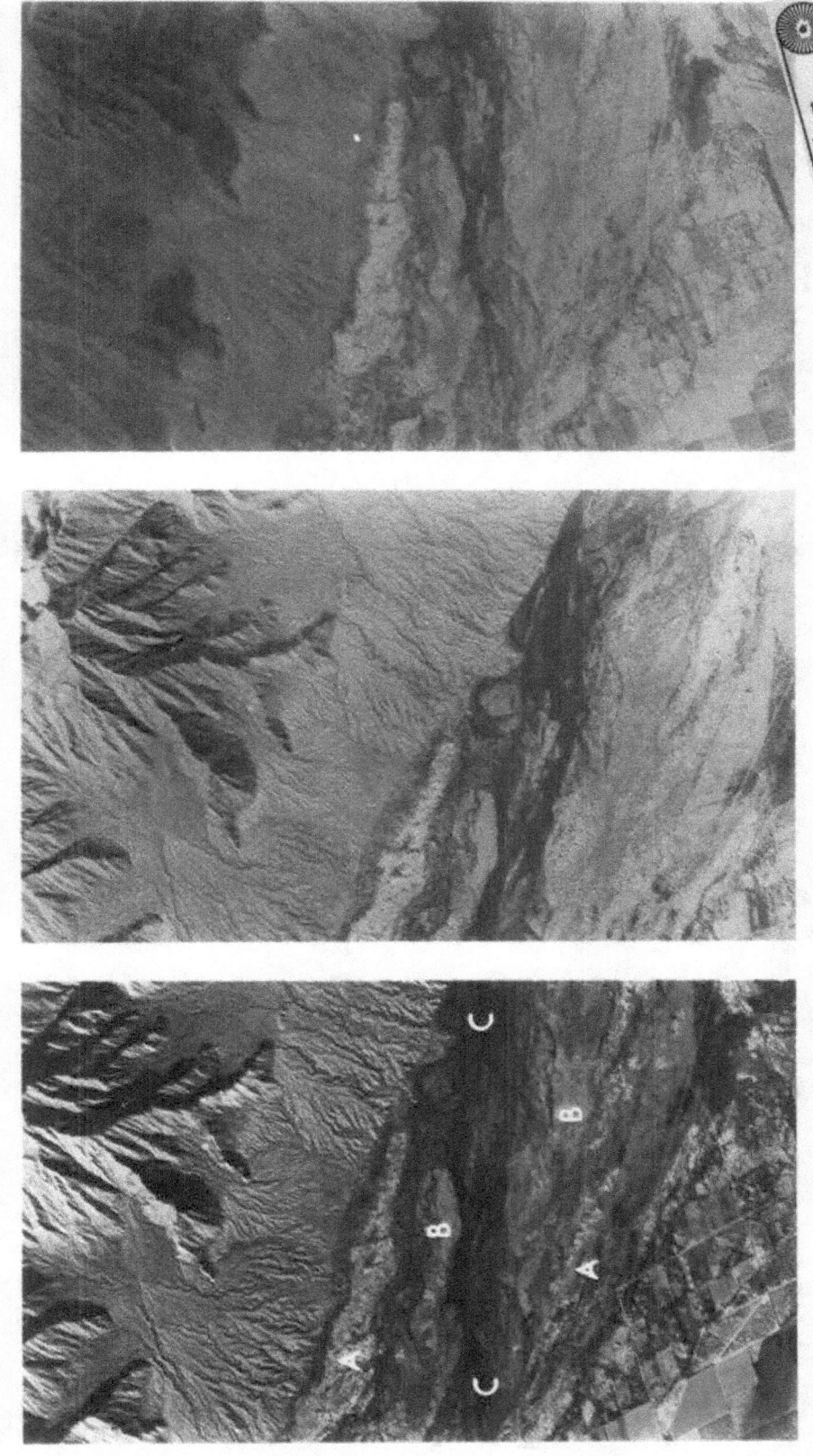

March 12, 1969   April 23, 1969   August 5, 1969

Color Plate 12. Infrared Ektachrome photos taken with a HyAc panoramic camera on the dates indicated. Three conspicuous terrain types within the river channel of the Gila River southwest of Phoenix are shown here. Indicated on the March 12 photo are salt flats (A), areas with annual grass cover (B), and stands of mesquite and tamarisk (C). See text for discussion.

Infrared Ektachrome

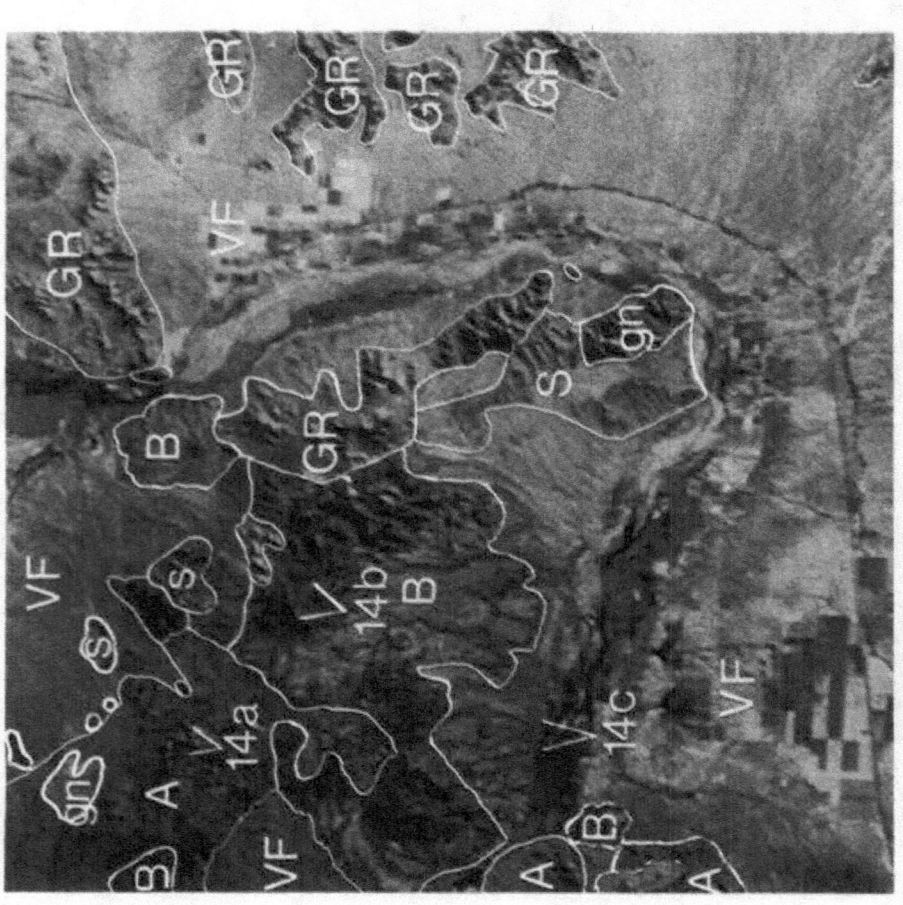

Ektachrome

Color Plate 13. Infrared Ektachrome (left) and Ektachrome (right) Apollo 9 photos of Gila Bend area. The geologic delineations on the Infrared Ektachrome photo were located with reference to the Arizona Bureau of Mines Geologic Map of Maricopa County and adjusted in conjunction with discernible geologic data from the photo. Rock types include igneous: basalt (B), andesite (A), and granite (GR); metamorphic: schist (s) and granitic gneiss (gn) rocks; and sedimentary (S) rocks. Most of the area is covered with unconsolidated valley fill (VF). Areas of valley fill can be readily identified by their topographic configuration, uniform tone, and parallel drainage patterns. Some features are more clearly displayed on the Ektachrome photo than on the Infrared Ektachrome photo. Alluvium in the low-water stream channel (lc) is more quickly contrasted with adjacent alluvium occupying the remaining portion of the channel (which is covered only in times of high water) on the Ektachrome print than on the Infrared Ektachrome print. Part of the difference in contrast, however, could also be a function of the difference in Sun angle; the Infrared Ektachrome having been taken at 8:30 a.m. (Sun elevation of 32°) and the Ektachrome an hour and a half later at 10 a.m. (Sun elevation of 48°). The areas 14a, 14b, and 14c refer to Color Plate 14.

(a)

(b)

Color Plate 14. (a) This oblique photograph of andesite area shows the reddish-brown tone and highly dissected nature of the low angular hills. The combination of this tone and topographic texture helps make the andesite distinctive from the basalt on the space photo (Color Plate 13). (b) In the upper right portion of this photograph are small granite hills that are clearly visible and mappable on the space photo. The light tone of the granite and its topographic form (generally long, then acutely branching ridges) distinguish it from the more basic extrusive igneous rocks present. In this area, however, it is interesting to note that the granitic gneiss cannot be easily differentiated from the granite due to similar tone and form. (c) The basalt in this photograph appears darker in tone than the andesite, and its topographic form is also different in that the basalt has been less eroded and forms longer, higher ridges or isolated conical hills. In extensive areas of flatter topography, the basalt is in evidence by the dark black gullies and scarp edges.

(c)

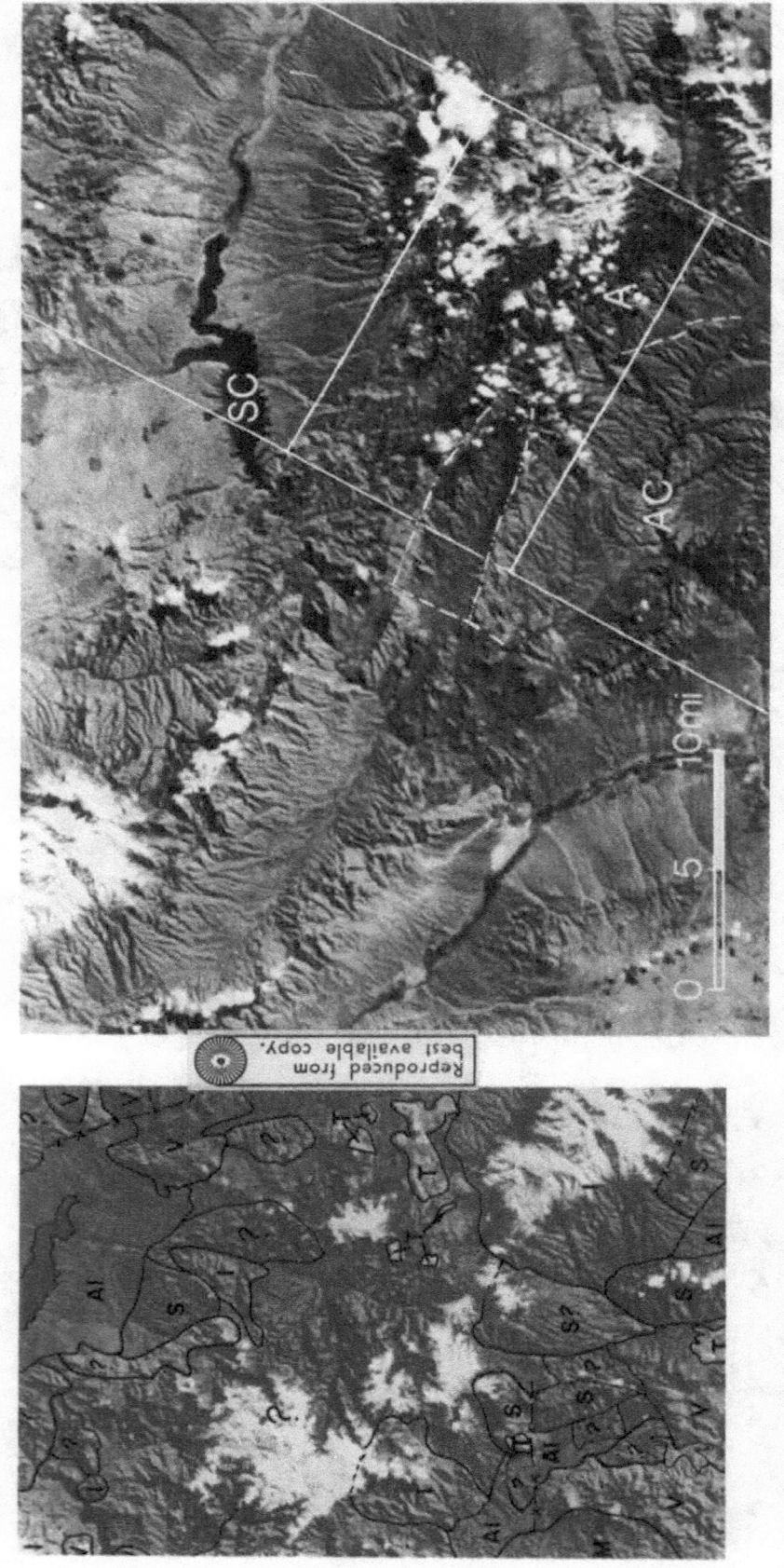

(a) This Infrared Ektachrome photo with geologic interpretation is for comparison and interpretation with the black-and-white bands of the S065 and with the ground data in figure 3–25.

(b) This 5½ × enlargement of a portion of an Apollo 9 transparency shows the flight strip covered by the high-altitude photography. The fault traces west of Aravaipa show up equally well on both the Apollo photo and the high-altitude photos, shown in Color Plate 16. However, the fold structure outlined by the overlay is more readily discernible and the structure more completely interpreted from the high-altitude photo in Color Plate 16. The Aravaipa Canyon region is also shown in figure 3–26.

Legend

A   Aravaipa (a town   AC   Aravaipa Canyon   ———— Fold structure
      southeast of Phoenix) SC   San Carlos Reservoir   – – – Possible fault

Color Plate 15

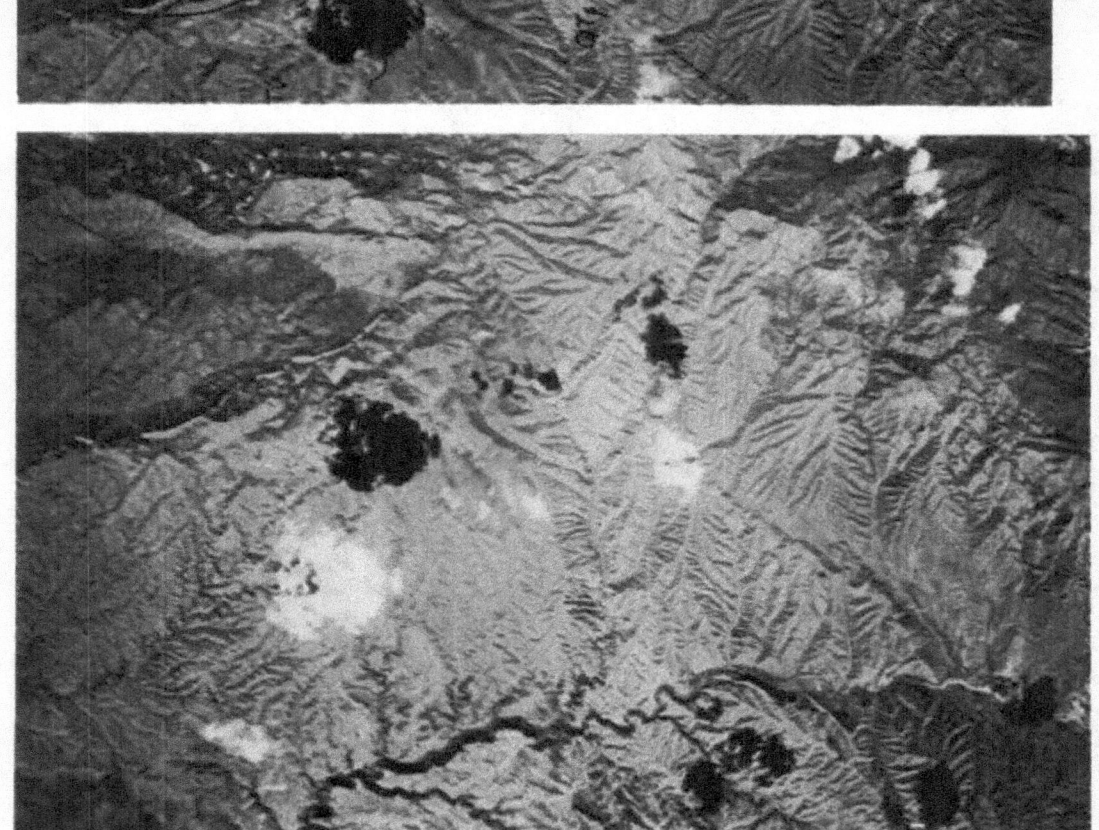
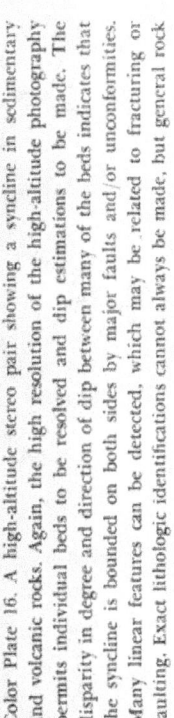

Legend

F   Faults or fractures

QTs   Quaternary-Tertiary sediments, loosely consolidated

S/V   Layered rock, i.e., sedimentary and/or volcanics

Tkv   Tertiary-Cretaceous volcanics

Tv   Tertiary volcanics

- - - -   Unit boundaries

....   Marker beds

———   Faults or unconformities

—·—·—   Linear features—possible faults or fractures

→   Direction of dip

Color Plate 16. A high-altitude stereo pair showing a syncline in sedimentary and volcanic rocks. Again, the high resolution of the high-altitude photography permits individual beds to be resolved and dip estimations to be made. The disparity in degree and direction of dip between many of the beds indicates that the syncline is bounded on both sides by major faults and/or unconformities. Many linear features can be detected, which may be related to fracturing or faulting. Exact lithologic identifications cannot always be made, but general rock categories can be determined. From the photo it can be said that the syncline occurs in layered rock that is sedimentary and/or volcanic in nature. The Quaternary-Tertiary sediments are fairly easily identified, and erosion patterns show they are not firmly consolidated. The crystalline mass in the lower part of the photo is fairly easily delineated, but from the photo it is not possible to tell whether it is igneous or metamorphic rock.

FIGURE 3–1.—Enlarged portion of an Apollo 9 Infrared Ektachrome photograph (AS9–26–3801) showing the location of test sites near Mesa, Ariz. The 16-sq-mi area (outlined at left) is the primary area for collection of ground truth and is the area used for administering tests for crop identification using high-altitude imagery from various dates. Enhanced images have also been prepared from multiband photographs of this area. The area outlined on the right was also monitored on the ground throughout the 1969 and 1970 growing seasons to provide additional information on crop type and variability, and for use as an "extended area" in future tests.

Ground photographs of representative fields have been taken to document how their conditions change throughout the growing season. Examples of this photography can be seen in Color Plate 4.

On April 23 stalks were beginning to appear in the barley field, the cotton field looked bare because seedlings were just emerging, and the alfalfa field looked dark because it had recently been mowed.

On May 21 the barley crop was mature, the cotton plants were still very young, and the alfalfa field had recently been mowed.

On August 5 the barley had been harvested and the field tilled, and foliage covered the cotton and alfalfa fields.

Low-altitude aerial oblique photographs were also taken over the study area in March and May of 1969. These two types of supporting photography were used to provide more detailed information on individual fields; they also became useful during the interpretation process when an attempt was made to explain subtle tone differences between fields, not only in terms of crop type but also in terms of the particular stage of crop growth and any transitory characteristic (irrigation, plowing, weed growth, etc.).

It was recognized at the outset that great difficulty might be encountered in attempting to identify, on early March photographs alone, the variety of crop types known to grow in the Phoenix area. Therefore an effort was made to use several aids to interpretation, including (1) sequential photographs, (2) tone-signature analysis, (3) multiband and multidate image enhancements, and (4) preliminary training exercises for the photointerpreter. Since sequential photographs, densitometric data, and image enhancements (color composites) were examined in the interpretation tests along with the four film-filter combinations from Apollo 9, they will be discussed before describing the test and the results.

Sequentially Obtained High-Altitude Photography

The possibilities for making Earth resource surveys on space photography cannot be completely evaluated until the merits of obtaining such photography sequentially (e.g., at weekly or monthly intervals) have been considered. It is in this sense that the monitoring of crops, by means of photography, becomes an especially important concept. Since it was known that it would not be possible to obtain sequential space photography (similar to that taken on the Apollo 9 mission) on succeeding dates throughout the 1969 growing season, sequential aerial photography was obtained during that period from an altitude of approximately 70 000 ft. This photography was obtained at intervals of about 1 month. It closely simulates the type and

18

Legend for Figures 3-2 and 3-16

A	Alfalfa	M_g	Milo (grain sorghum); green, reaching maturity
A_c	Alfalfa recently cut and alfalfa pastures		
A_m	Alfalfa nearing maturity or ready to be cut	M_m	Milo (grain sorghum); mature, ready for harvest
A_s	Alfalfa recently seeded	OA	Oats
B	Barley	OA_c	Oats; field harvested
B_o	Barley; field harvested, stubble remaining	OA_g	Oats; young plants developing
B_g	Barley; young plants developing	P	Pasture
B_i	Barley; field green, inflorescence emerged	S	Forage sorghum
B_m	Barley; mature, ready for harvest	S_c	Forage sorghum; crop harvested
BS	Cultivated fields not yet planted	S_g	Forage sorghum; plants green and vigorous
BS_d	Cultivated fields not yet planted (bare soil relatively dry)	SB	Sugar beets; foliage cover consists of large beet leaves combined with weeds and scattered cereal plants from previous continuous cover crops
BS_m	Cultivated fields not yet planted (bare soil relatively moist from recent irrigation)		
C	Cotton	STUB	Stubble
C_g	Cotton; plants reaching maturity, foliage green	W	Wheat
C_h	Cotton; crop harvested	W_o	Wheat; harvested, stubble remaining
C_m	Cotton; plants mature, defoliated	W_g	Wheat; young plants developing
CO	Corn	W_i	Wheat; field green, inflorescence emerging
DISC	Disked	W_m	Wheat; mature, ready for harvest
F	Fallow field	U. of A. Exp. Farm	University of Arizona Experimental Farm
H	Housing development or other structures		
M	Milo (grain sorghum)		
M_o	Milo (grain sorghum); field harvested		

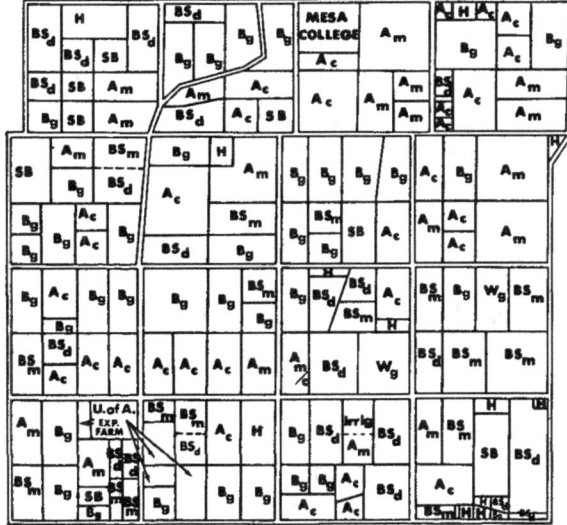

FIGURE 3-2.—Ground data map indicating crop type and condition, Mesa test site, March 12, 1969. At this time the B_g fields are green, 18-in. average height, inflorescence not emerged. The W_g fields are green, 16-in. average height, inflorescence not emerged.

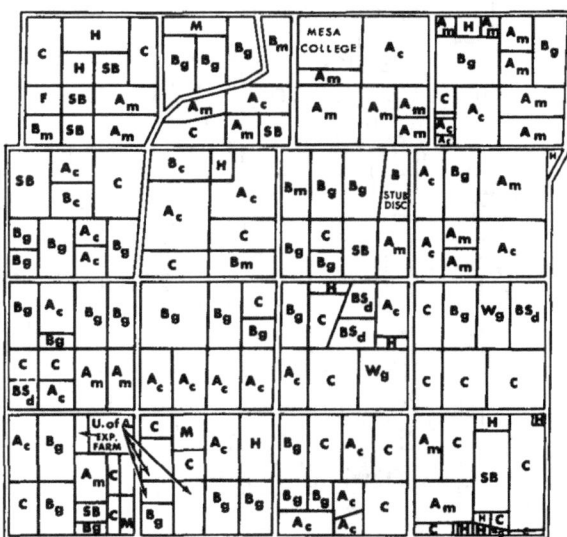

FIGURE 3-3.—Ground data map indicating crop type and condition, Mesa test site, April 23, 1969. The B_g and W_g fields are still green and reaching maturity; cotton and milo seedlings are just emerging.

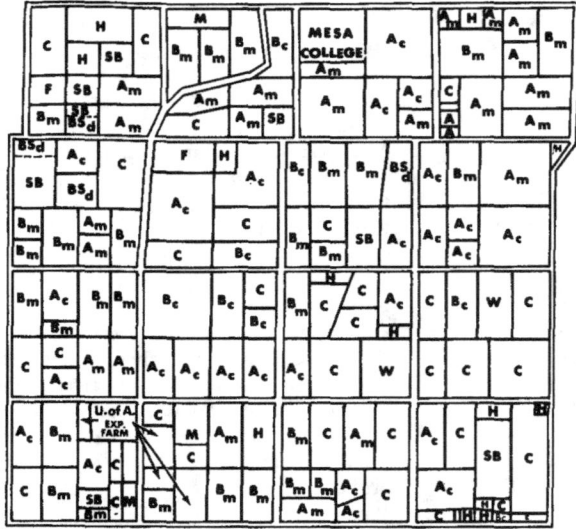

FIGURE 3–4.—Ground data map indicating crop type and condition, Mesa test site, May 21, 1969. The wheat fields are reaching maturity; some sugar-beet fields have been harvested; young cotton and milo plants are emerging.

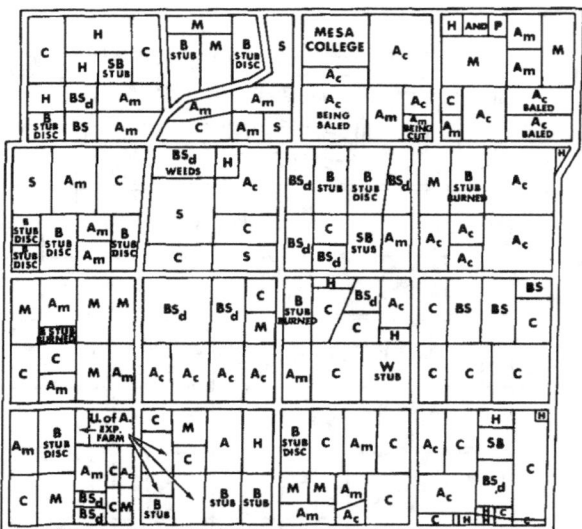

FIGURE 3–5.—Ground data map indicating crop type and condition, Mesa test site, July 15, 1969. Some sugar-beet fields have been harvested.

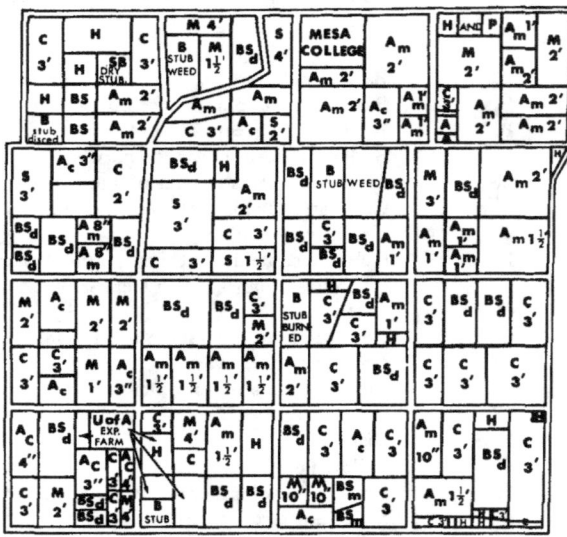

FIGURE 3–6.—Ground data map indicating crop type and condition, Mesa test site, August 5, 1969.

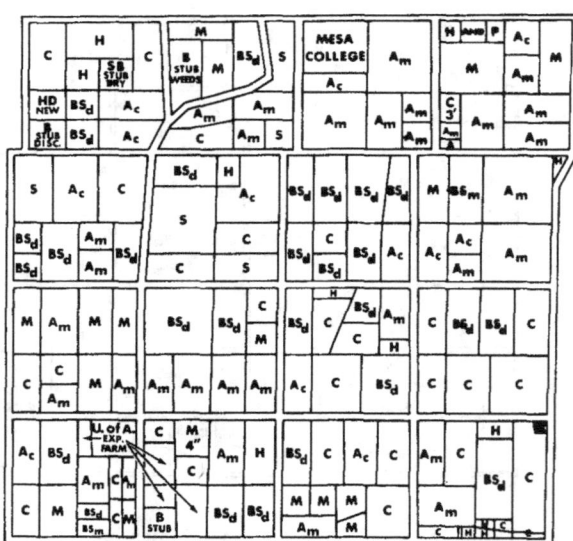

FIGURE 3–7.—Ground data map indicating crop type and condition, Mesa test site, August 29, 1969. All of the forage sorghum fields have green and vigorous plants.

20

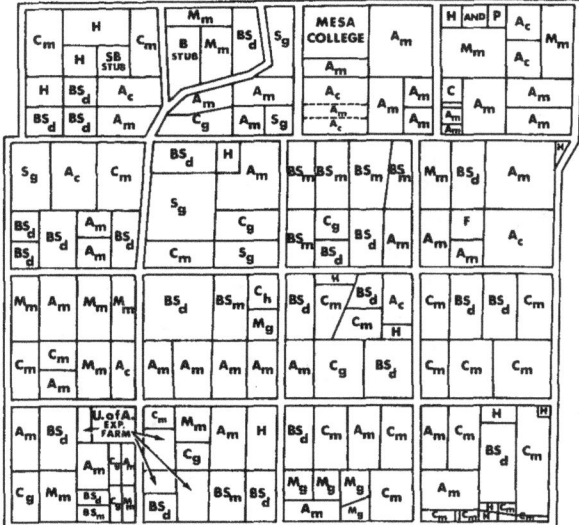

FIGURE 3-8.—Ground data map indicating crop type and condition, Mesa test site, September 30, 1969.

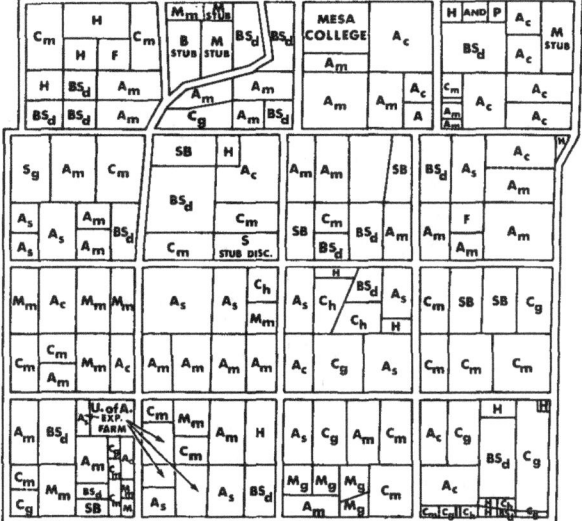

FIGURE 3-9.—Ground data map indicating crop type and condition, Mesa test site, November 4, 1969. Note that in this month the sugar beets are all young plants.

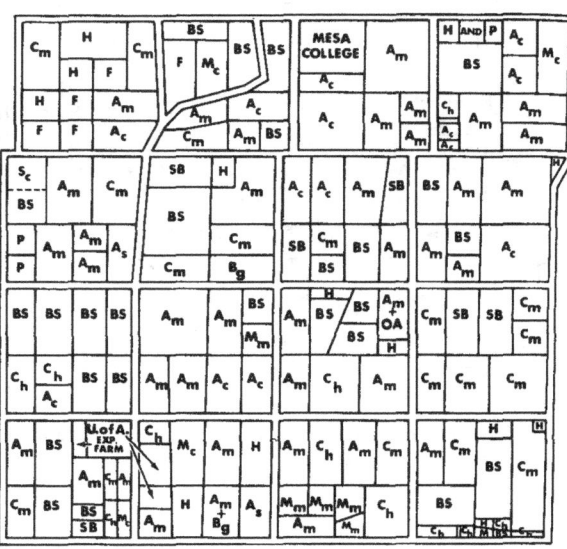

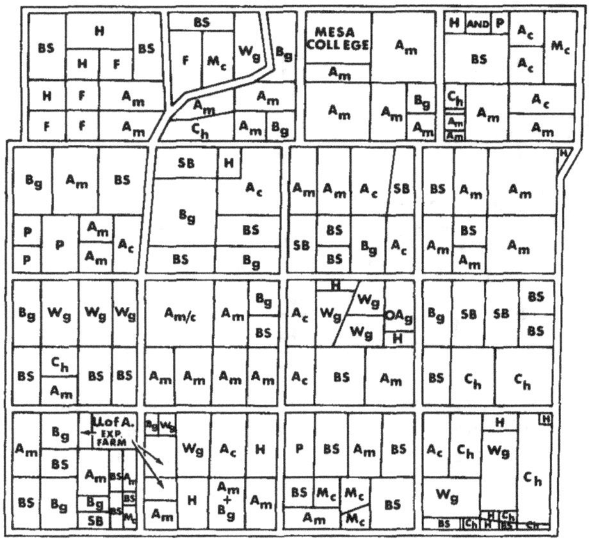

FIGURE 3-10.—Ground data map indicating crop type and condition, Mesa test site, December 9, 1969.

FIGURE 3-11.—Ground data map indicating crop type and condition, Mesa test site, January 18, 1970.

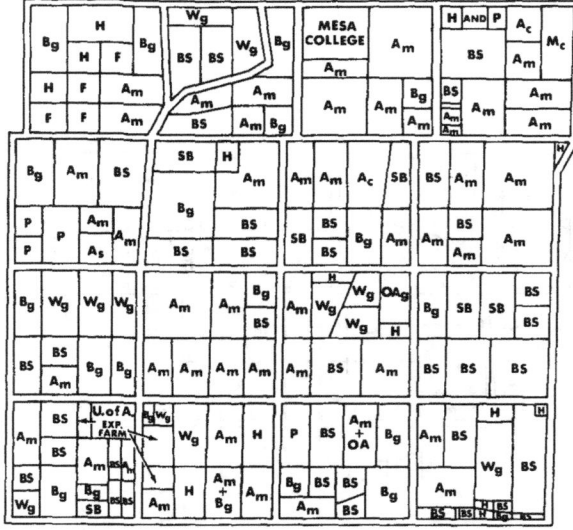

FIGURE 3–12.—Ground data map indicating crop type and condition, Mesa test site, February 6, 1970.

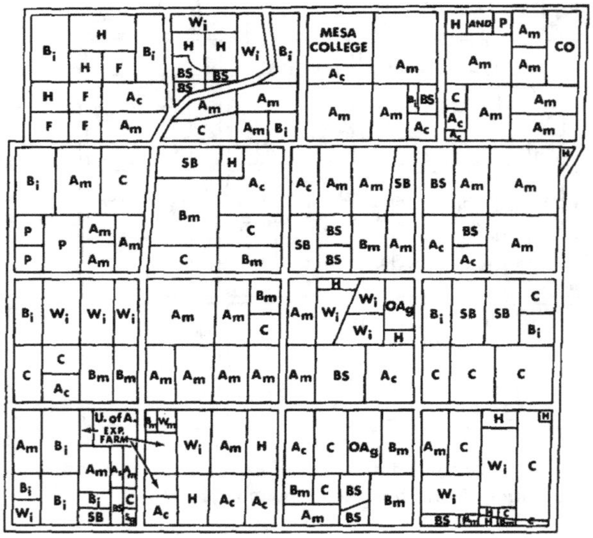

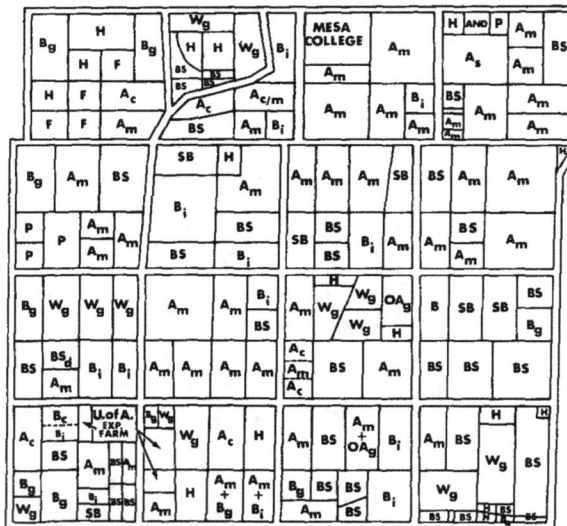

FIGURE 3–13.—Ground data map indicating crop type and condition, Mesa test site, March 15, 1970. Note that in all the OA$_g$ fields the inflorescence has not emerged.

FIGURE 3–14.—Ground data map indicating crop type and condition, Mesa test site, April 17, 1970. In the OA$_g$ fields the inflorescence has not emerged; the S$_g$ fields have seedlings just emerging.

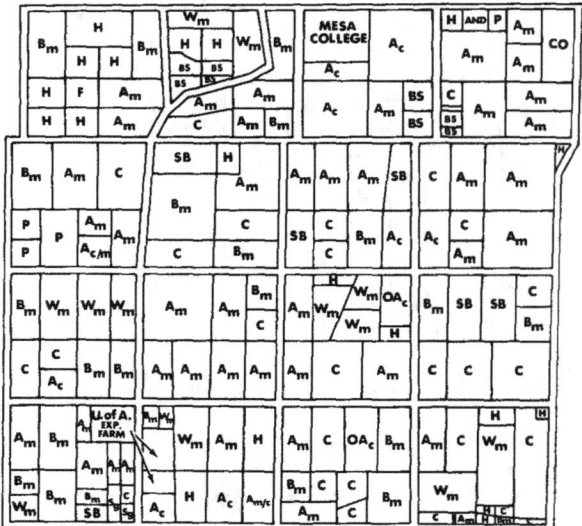

FIGURE 3–15.—Ground data map indicating crop type and condition, Mesa test site, May 21, 1970. The cotton plants are just emerging; young plants are developing in all of the CO and S$_g$ fields.

FIGURE 3–16.—Ground data map indicating crop type and condition, Mesa test site, June 11, 1970. The cotton plants have grown to 6 to 12 in.; the forage sorghum is in the grass stage.

quality of imagery expected from space on a sequential basis within a few years.

Of all the Earth resource features that might be investigated sequentially, agricultural croplands are among those exhibiting the most dramatic changes within a relatively short time. Thus our initial effort in studying sequential photography of the area centered around Phoenix has been oriented toward determining how much more information about agricultural resources could be derived from it.

High-altitude multiband photography of the Mesa agricultural test site was acquired on March 12, 1969, concurrent with the flight of Apollo 9. This imagery was obtained from an altitude of 70 000 ft giving a photo scale of 1/250 000, using four Vinten cameras (70-mm format). In addition, two HyAc panoramic cameras containing Infrared Ektachrome and Panchromatic film were part of the camera system.

Similar aerial photographs were obtained during 1969 on April 23, May 21, August 5, September 30, and November 4 (Color Plate 5, fig. 3–17). Four Nikon cameras replaced the Vinten cameras for these subsequent missions. The photographs from the Nikon cameras are 35-mm format (21-mm focal length) giving a photographic scale of ap-

proximately 1/850 000. Continuing coverage of the Phoenix area on small-scale aerial photographs was obtained from an altitude of 60 000 ft using the NASA-based RB57F jet aircraft from December 1969 to June 1970. Data have been obtained using two RC–8 cameras (equipped with Ektachrome and Infrared Ektachrome film and providing a scale of approximately 1/120 000) and four Hasselblad cameras equipped with Ektachrome, Infrared Ektachrome, and black-and-white films, providing a scale of approximately 1/450 000.

Factors Affecting Crop Identification

Once data of the kind just described have been obtained, techniques for identifying crop type can be developed. The most serious limitation to developing useful techniques lies in the variability of crop type and cropping practices. Any factor which affects the distribution, development, and vigor of a crop may affect its photographic signature, and thus may influence the success with which that crop can be consistently identified.

Crop Type and Distribution

Agricultural practices in an area are generally stable. Totally foreign crops are rarely introduced. For this reason, interpretation keys can be devised for a specific area with little fear that certain crops will totally disappear or that new crops will suddenly be introduced on a large scale. Table 3–1 shows that these generalizations are valid for the Arizona area and for the Nation as a whole as applied to the main crops grown in Arizona during a recent 4-yr period. Changes in acreage are generally small, although the data for sugar beets point up an interesting exception. Sugar beets were only introduced in Arizona in 1966 when allotments for the planting of this crop were granted by the Federal Government, and in 1968 and 1969 acreage rose substantially with governmental approval. It is probable that the sugar-beet acreage of Arizona will expand in the years to come. In using sequential photography year after year as an aid to crop identification, one must consider such trends.

The geographic distribution of crop types and the degree of similarity of cropping patterns from one area to another are factors that govern the validity of an interpretation key for identifying crops in widely separated regions. Crop identifica-

TABLE 3–1.—*Acreage Planted to Various Crops in Arizona as Compared with U.S. Acreage, 1967–1969*

[Source: Arizona Crop and Livestock Reporting Service, 1969[a]]

Crop	Acres planted in Arizona, by years [a]					Acres planted in the United States, by years [a]				
	1967	1968	Indicated [b] 1969	Actual [b] 1969	Actual 1969 as percent of 1968	1967	1968	Indicated [b] 1969	Actual [b] 1969	Actual 1969 as percent of 1968
Cotton	248	298	320	312	104	9 448	10 921	12 012	11 898	108
Barley	187	200	168	166	83	10 002	10 322	10 352	10 158	98
Sorghums	254	239	206	208	87	19 007	17 924	17 659	17 438	97
Suga beets	13.2	19.6	34.0	33.6	171	1 197	1 481	1 649	1 650	111
All hay	233	239	225	224	93	64 667	62 570	62 730	61 838	98

[a] In thousands of acres.

[b] "Indicated 1969" values reflect predictions by farmers as to their intentions to plant; "Actual 1969" values are acreages actually planted.

tion characteristics in the Mesa test site in Arizona have been compared with those in the Imperial Valley test site of California (ch. 4). The results obtained suggest that there is enough similarity to permit common photointerpretation keys and techniques to be used for identifying many of the same crops in both areas.

Seasonal Development

Documentation of the seasonal development of crops is important for determination of optimum times of the year for crop discrimination. Both within-season and between-season variability will affect the specification of optimum dates for obtaining photography. For the 16-sq-mi test site near Mesa, field-by-field tabulations of crop type and condition were made from the ground data collected on each mission date. This information was used to write the generalized descriptions of crop development in table 3–2. From this information, the calendar in figure 3–18 was prepared for the major crop categories within the 16-sq-mi Mesa test site (Maricopa County). Included in the calendar are descriptions of the seasonal changes that occurred within each crop type during the 1969 growing season.

Within the 1969 growing season, four distinctive crop sequences were noted. These include:

(1) Cereal grains (wheat and barley) followed by sorghum (grain or forage)

(2) Bare soil followed by cotton

(3) Mature alfalfa alternating with freshly cut alfalfa (generally mowed monthly)

(4) Sugar beets followed by grain sorghum (milo) or bare soil

These sequences can be explained in terms of the growth patterns of each crop type. For example, cotton must be planted in early spring (more than a month earlier than sorghum); hence, it can be found only in fields that appeared bare on March photographs, it cannot follow a cereal grain or sugar-beet crop. Sorghum, on the other hand, can be planted in late June or July and still reach maturity before the end of the growing season. Thus, it can be planted in fields that contained barley or sugar beets, once these crops have been harvested in May and June. Alfalfa follows an entirely different pattern, because it is grown in the same field for more than 1 yr (generally 2 or 3 yr). It is mowed at monthly intervals and its appearance changes on the photographs depending upon the time since the last mowing. Representative ground photographs illustrating these points can be found in Color Plate 4.

Comparison of data for 1969 and average data for crop development over other years shows that patterns for 1969 were generally normal (Arizona Crop and Livestock Reporting Service, 1969*b*). Cotton harvesting was delayed because of late boll set and wet weather (only 60 percent of the crop had been picked and ginned by December 1, 1969, as opposed to 84 and 85 percent in 1968 and 1967, respectively). Sugar-beet yield was affected by heavy weed development in the fields and by insect attacks. Also, late-planted grain sorghum fields were of poor quality and harvest was delayed by

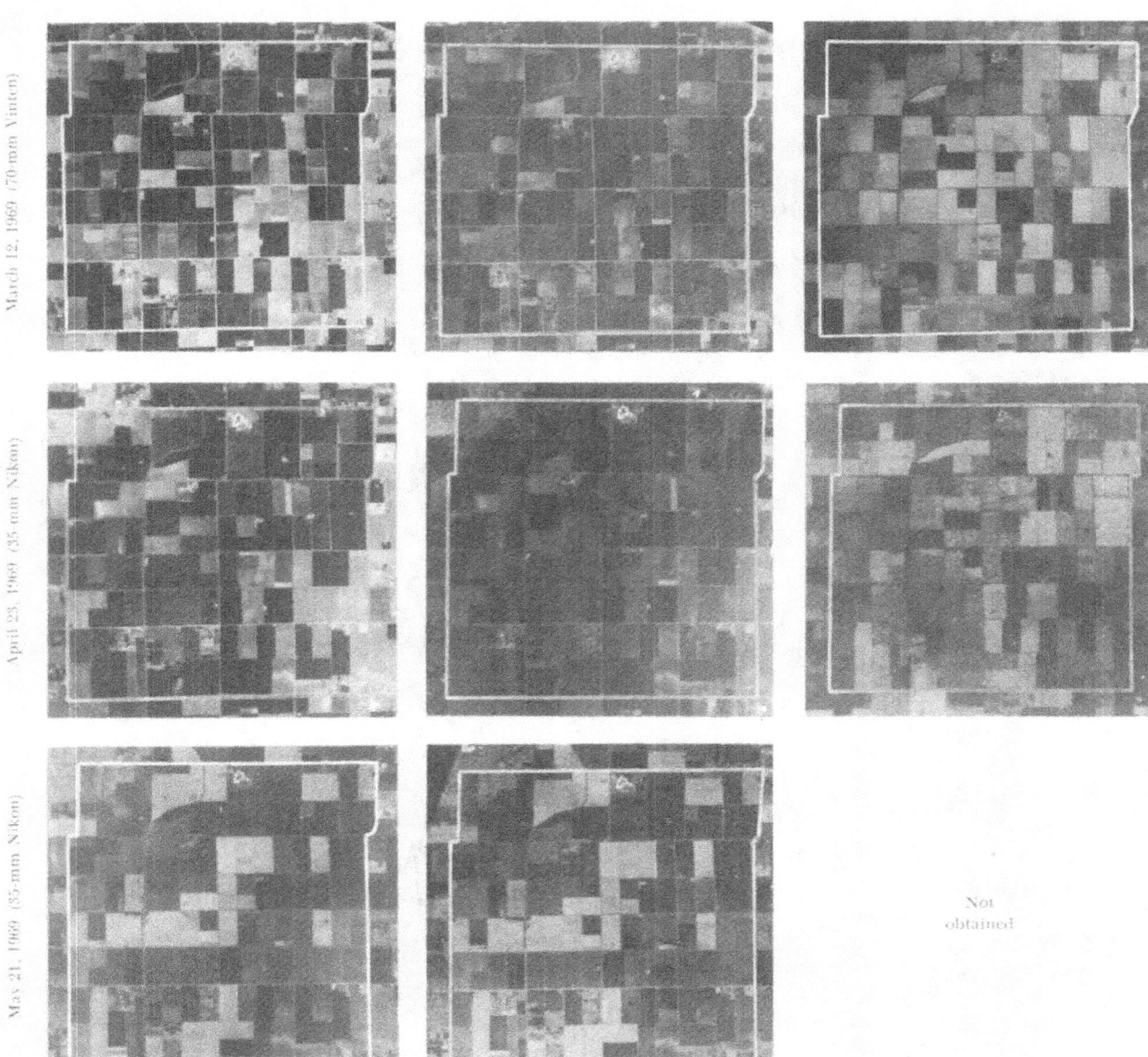

Pan-25A   Pan-58   IR–89B

March 12, 1969 (70-mm Vinten)

April 23, 1969 (35-mm Nikon)

May 21, 1969 (35-mm Nikon)

Not
obtained

FIGURE 3–17. — Multiband black-and-white photographs (Pan-25A, Pan-58, and IR–89B) of the 16-sq mi Mesa test site (outlined) taken from an altitude of approximately 70 000 ft on the dates indicated. These prints are copies from the 35-mm Nikon images for all dates except March 12 when a Vinten 70-mm-format camera system was used. All images as reproduced here have been enlarged to a common scale of 1/130 000. Infrared 89B photographs are missing for two dates, May 21 (camera malfunction) and August 5 (extreme overexposure). The poor quality of the other August 5 images is a result of cirrus clouds.

Pan-25A   Pan-58   IR-89B

August 5, 1969 (35-mm Nikon)

September 30, 1969 (35-mm Nikon)

November 4, 1969 (35-mm Nikon)

Image unusable

TABLE 3–2.—*Summary of Crop Development for*

| Crop type | Crop development by date of photography— | | | |
	March 12	April 23	May 21	July 1
Wheat, barley	Vigorous growth, green foliage; 100 percent cover; inflorescence about to emerge.	Some fields maturing, others green; 100 percent cover; inflorescence developed.	Fields matured and dry, many harvested; stubble plowed in some cases.	Fields of stubble or tilled (bare) soil.
Alfalfa	(a)			
Sugar beets	Green foliage; 100 percent cover.	Green foliage; 100 percent cover.	Some fields harvested, tilled; others green with 100 percent cover.	Most fields harvested, residual leaves, bare soil, or emerging sorghum.
Cotton	Fields bare, prepared for planting.	Seedlings emerging; less than 1 percent cover.	Young plants 6 to 12 in.; less than 5 percent cover.	Plants green, 2 to 3 ft tall; cover 80 to 100 percent.
Milo (grain sorghum)	---		Milo planted in fields previously occupied by barley, sugar beets, or bare soil.	Seedlings less than 6 in. emerging; less than 5 percent cover.

NOTE.—The information in this table was used to construct the crop calendar in fig. 3–18.

a Variable height and density of leafy green foliage, depending upon time interval since last cutting; fields cut at approximately 1-month intervals, leaving coarse stems.

wet field conditions. In general, the pattern was near normal; thus, conclusions reached regarding the optimum time for discrimination of particular crops made this year should hold for future years if they follow the established trends.

Comparisons of crop data from one year to another were also made for the 16-sq-mi Mesa test site. Information regarding crop type in each field was compared for 1969 and 1970. The data in table 3–3 were obtained.

These data suggest that there are some definite patterns of crop sequence from one year to another. For example, in 1970, sugar beets were planted only in fields which contained a cereal grain (wheat or barley) in 1969. Also, cereal grains only (no other crop) were grown in fields which produced sugar beets in 1969. Alfalfa continued to grow in 36 fields out of 50 that contained that crop in 1969. Alfalfa is not an annual crop (one in which the plants grow to maturity, die, and are removed each year). Instead, the same alfalfa plants are grown for 2 or 3 yr in a given field.

Further search for crop patterns showed that a second cotton crop was grown in 55 percent of the fields in which it grew during 1969, and a cereal grain crop was grown in all but two of the remaining 14 fields that contained cotton in 1969. Also, seven fields in this 16-sq-mi test area were taken from agricultural production and converted to subdivisions. The expansion of the city of Tempe was responsible for this shift in land use. More of this conversion can be expected as the population in the area expands.

The generalizations which have just been drawn

the Mesa Test Site During the 1969 Growing Season

| Crop | Crop development by date of photography—Continued | | | | |
	July 15	August 5	August 29	September 30	November 4
Wheat, barley	Fields tilled, bare (few remain with stubble).	---------------------------------			Fields being tilled and prepared for planting 1970 crop.
Alfalfa	(ª)				
Sugar beets	Residual leaves, bare soil, or young sorghum.	---------------	Fields bare, preparations underway for planting of 1970 crop.	Planting of new crop complete.	Green foliage; 10 to 50 percent cover.
Cotton	Plants green, 2 to 3 ft tall; cover 90 to 100 percent.	Plants green, 3 to 4 ft tall; some fields flowering; cover 90 to 100 percent.	Plants green, 3 to 4 ft tall; cotton bolls mature on some plants, flowers developing on others; cover 90 to 100 percent.	Plants 3 to 4 ft tall, foliage dry, cotton bolls mature; cover 90 to 100 percent.	Fields mature; many fields have been harvested (plant residue remaining).
Milo (grain sorghum)	Plants developing (12 to 24 in. tall), inflorescence emerging; percent cover depends on stage of maturity (10 to 100 percent).	Inflorescence maturing; 100 percent cover.	Plants 3 ft tall; 100 percent cover; inflorescences maturing; some fields ready for harvest.	Plants 3 to 4 ft tall; 100 percent cover; inflorescences mature; fields ready for harvest.	Many fields harvested (stubble remaining or disked); other fields mature, ready for harvest.

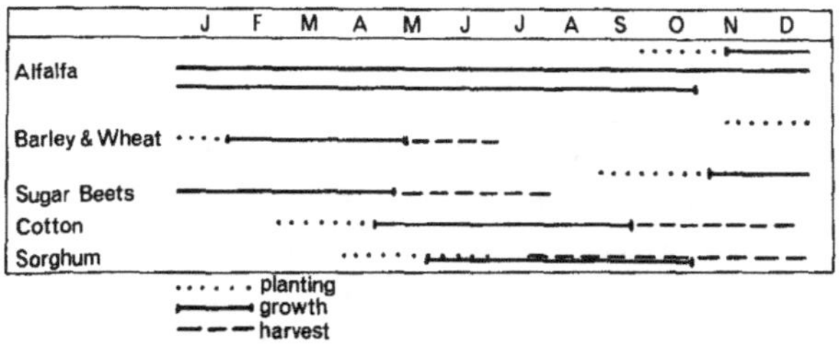

FIGURE 3–18.—This crop calendar summarizes the development patterns of five major crop types in the Maricopa County test site. The duration of each of the three main phases of development (planting, growth, and harvest) is indicated. It was prepared using field data and published crop status reports for Maricopa County. This kind of information is used to select optimum dates for discrimination of each crop type on aerial and space photography.

are useful in predicting the identity of crops in future years and may help to improve the accuracy of crop surveys.

Knowledge of crop sequences and of the variations that affect these sequences is needed to specify the optimum times of the year for obtaining sequential photography. For agricultural areas, the cyclic changes are best summarized in a crop calendar such as that shown in figure 3–18. Tone values of individual fields (as seen on photographs of a given date) can be related to the stage of maturity of the crops on that date, as summarized in the crop calendar. The calendar can be used to determine either (1) when a particular crop type exhibits a condition that produces a unique signature that can be discriminated from signatures of all other crops, or (2) what combination of dates for sequential photography is best for crop identification.

Barley and cotton can be identified easily by the first of these techniques. Barley is the only crop in

TABLE 3–3.—*Crop Sequences, 1969 and 1970*

Spring 1969		Spring 1970		
Crop	No. of fields	Crop	No. of fields	Percent of 1969
B_____	35	B_____	6	17 ⎫ 26
		W_____	3	9 ⎭
		A_____	14	40
		C_____	4	11
		SB_____	5	14
		P_____	3	9
W_____	2	A_____	1	50
		SB_____	1	50
A_____	50	A_____	36	72
		C_____	3	6
		S_____	3	6
		B_____	3	6 ⎫ 10
		W_____	2	4 ⎭
		Other_____	3	6
C_____	31	C_____	17	55
		B_____	8	26 ⎫ 39
		W_____	4	13 ⎭
		Other_____	2	6
SB_____	5	B_____	4	80 ⎫ 100
		W_____	1	20 ⎭
SB, B, BS_____	7	H_____	7	

NOTE.—The crop symbols are the same as those used in figures 3–2 to 3–16.

the 16-sq-mi test site that has a characteristic golden color when mature in May (Color Plate 4). Since barley has a characteristic color on the Infrared Ektachrome photograph of May 21 (see Color Plate 5), it can be distinguished from other crops with a high degree of success. (See later section for results from interpretation tests.) The unique cropping pattern for cotton also aids in its identification. Because cotton is planted only in fields that are bare in March, the location of the cotton crop can be predicted in March by identifying bare soil fields. Later in the year, the presence of cotton can be confirmed.

Other studies have shown that no single date is best for making identification of all important features, but that data from each month can profitably be used for specific purposes. Sayn-Wittgenstein (1967) suggests that *winter* photographs will permit open plains to be distinguished from wooded hills by their sharp contrast at that season; *spring* photographs are best used for studying the ground surface and for differentiation of

hardwood and conifer tree species; and photographs taken during the *fall* are best used to evaluate topography and identify soils. To select the best date for procuring aerial photography, Sayn-Wittgenstein notes that it is essential to understand the seasonal effects that govern the appearance of the ground features being studied. Brunnschweiler (1957), working in Switzerland with 1/12 000-scale photographs, found that image contrast was greatest on May and August photographs and that problems in field structure (size, shape, and boundaries of fields) could best be solved using photographs taken during those months. Individual crops could best be discriminated on June or July exposures. May photographs were judged best for interpretation of soil conditions. As in other studies of large-scale photographs (Goodman, 1959, 1964; Schepis, 1968; Steiner, 1969), dependence was made upon certain kinds of photographic detail (crop height, crop structure, cultivation and harvesting patterns, and type and location of farm buildings and equipment, for example) that cannot be detected on the high-altitude photographs of the type examined in our present study. For this reason, other parameters must be employed to provide useful information from very-small-scale images. Crop signature is one of the most important of these parameters.

Crop Signature

Because little field detail is discernible at the scale and resolution of the high-altitude Nikon photographs that were studied, photographic tone or color becomes the critical factor for identification. Either spectral signatures must exist at one date so that individual crop type can be identified, or else sequential patterns of tone or color must exist so that a crop type can be distinguished on the basis of changing patterns (i.e., bare soil to continuous cover crop to bare soil) at particular dates.

One of the most serious problems experienced during interpretation of the 1969 high-altitude photography resulted from the variability in image quality from one date to the next. This situation is evident in Color Plate 5 where a wide range of color balance exists among the photographs taken on different dates. The poor quality of the August 5 image was caused by high cirrus clouds that partially obscured the Phoenix area on that date.

However, the weather was clear on all other dates, so the differences in color balance then should be ascribed to exposure, film (age, storage, etc.), or processing differences. The prints reproduced in Color Plate 5 were copied from duplicate transparencies made from the original transparencies.

Sometimes the identification of crop types from sequential photographs depends merely upon whether fields are vegetated or nonvegetated (bare soil) at certain times of the year. In such instances the presence or absence of some red tone on Infrared Ektachrome photographs can be easily noted, regardless of the color balance of the images, and the problem of variability in image quality is not serious. When the discrimination between crops depends upon recognition of subtle color differences on Infrared Ektachrome photography, however, consistency in the color balance of images procured on different dates becomes important. Unless there is some consistent color signature on different dates for a crop, indicating that the crop is undergoing unique changes, sequential photointerpretation keys that use color as a dependable characteristic for crop identification cannot be prepared.

The preparation of color composite images from multiband photographs has been suggested as a means of reducing variability of color images taken on different dates. It has been theorized that adjustments could be made with a color combiner to insure that the color balance of selected images would match calibration targets. Presumably this would insure uniform color balance on the color composites, despite variations in gamma and related tonal qualities on multiband black-and-white photographs. This may be true if the composite images as formed on the screen are to be viewed directly. However, our tests show that if color reproductions of these composite images are to be made, as is often the case, the copy film and prints made from it must also be properly exposed and processed. In that event, some of the same problems that exist when seeking uniform quality in the original aerial color photography will be found to exist in color reproductions of the composite images. Figure 3–17 contains examples of multiband photographs from the high-altitude flights. All were copied in the same fashion from the duplicate rolls, and these prints indicate the general

quality and variability obtained. Note also that no IR–89B imagery was obtained for May 21 and August 5, thus limiting any possibility for using an infrared image in the preparation of enhanced images (e.g., simulated Infrared Ektachrome color composites) for those dates. Attempts were made in our studies to prepare color composites at uniform color balance for dates in which three multiband images were procured. However, the FRSL color combiner, which was used for this purpose, has limited capability for adjustment of brightness; and little success was achieved in preparing images of uniform color balance from one date to another. Also, since the data were not calibrated, there was no reference against which to standardize the enhancements.

Tone-Signature Analysis

As a means of investigating tone signature for the crop types found in the Phoenix area, measurements of light transmission on Pan-25A negatives for representative fields in the 16-sq-mi test site near Mesa were made. In this way the variability of tone signature for each crop type could be determined and compared. The relative success that might be achieved in distinguishing one crop from another could also be assessed by studying the relative differences in density among crop types.

A Welch Densichron instrument was used to measure negative transmission on Pan-25A negatives for March 12, April 23, May 21, August 5, and September 30. Several fields of each crop type (fig. 3–19) were measured on each negative, in order to include the range of transmission values of each crop type. These ranges are plotted in figure 3–20 for each image. Comparisons of data from a single image can be used to predict the probability of identifying particular crop types. For example, if the range of transmission values for one crop does not overlap the values for any other crop, a photointerpreter probably could identify that crop consistently. However, comparisons *between* images are not possible because no calibration information is available and no reference targets were established in the test sites on each date of photography.

The data in figure 3–20 indicate that there are very few cases in which one crop has a unique tone signature at a single date. Three examples can be

FIGURE 3–19.—These maps show the agricultural field patterns in the Mesa test site. The numbers associated with many of the fields indicate the percent light transmission of the Pan-25A negative for that particular crop type in March (*left*) and May (*right*). Comparison of these values gives an indication of the crop types which can be differentiated. (See the legend for figs. 3–2 to 3–16.) Sequential photographs permit an interpreter to increase the correct identification of crop categories. Barley fields, for example, are consistently identified by their characteristic signature (light tone) as seen on May photographs (plants mature). In March, however, barley fields appear similar to alfalfa (mature), wheat, and sugar beets; hence, the percent correct identification for barley is relatively low. BS¹ indicates fields now fallow or cultivated that had produced a barley crop in March. B² indicates barley fields not detected on May photographs because they contained oats that were still green.

discussed, however, that demonstrate a high correlation between transmission data and interpretation accuracy. Fields that contained dry bare soil in March have transmission values that do not overlap with any other field type. These fields have the lowest transmission values of any type category measured. (And they have the brightest tones on a positive print; see Pan-25A print for March 12 in fig. 3–17.) Interpretation results for dry bare soil (see table 3–16, image 3) indicate that this condition was identified correctly 87 percent of the time, with 41 percent commission errors. (See "Interpretation Tests" following for discussion of percent correct identifications and commission errors in interpretations.) In early April, a cotton crop was planted in each field that had been bare in March. By April 23, when high-altitude photographs were again obtained, the cotton seedlings were just emerging (Color Plate 4) and these fields appeared as bare soil. As in March, the transmission values

for this condition as measured on the April 23 negative were different from those for any other crop type (see graph for April 23 in fig. 3–20). Interpreters achieved 97 percent accuracy for identifying this category and made only 16 percent commission errors. This highly favorable result might have been expected from an examination of the light-transmission data of figure 3–20.

The record is similar for the identification of barley. For March 12 and April 23 negatives, transmission values for barley overlap those of several other crops, suggesting that it would be difficult to distinguish barley on either of these dates. This prediction is borne out by interpretation results. (See table 3–16.) Only 34 percent of the barley fields were identified correctly in March with 38 percent commission errors, and only 31 percent were identified correctly in April with 44 percent commission errors. The golden color of mature barley (Color Plate 4) produces a light

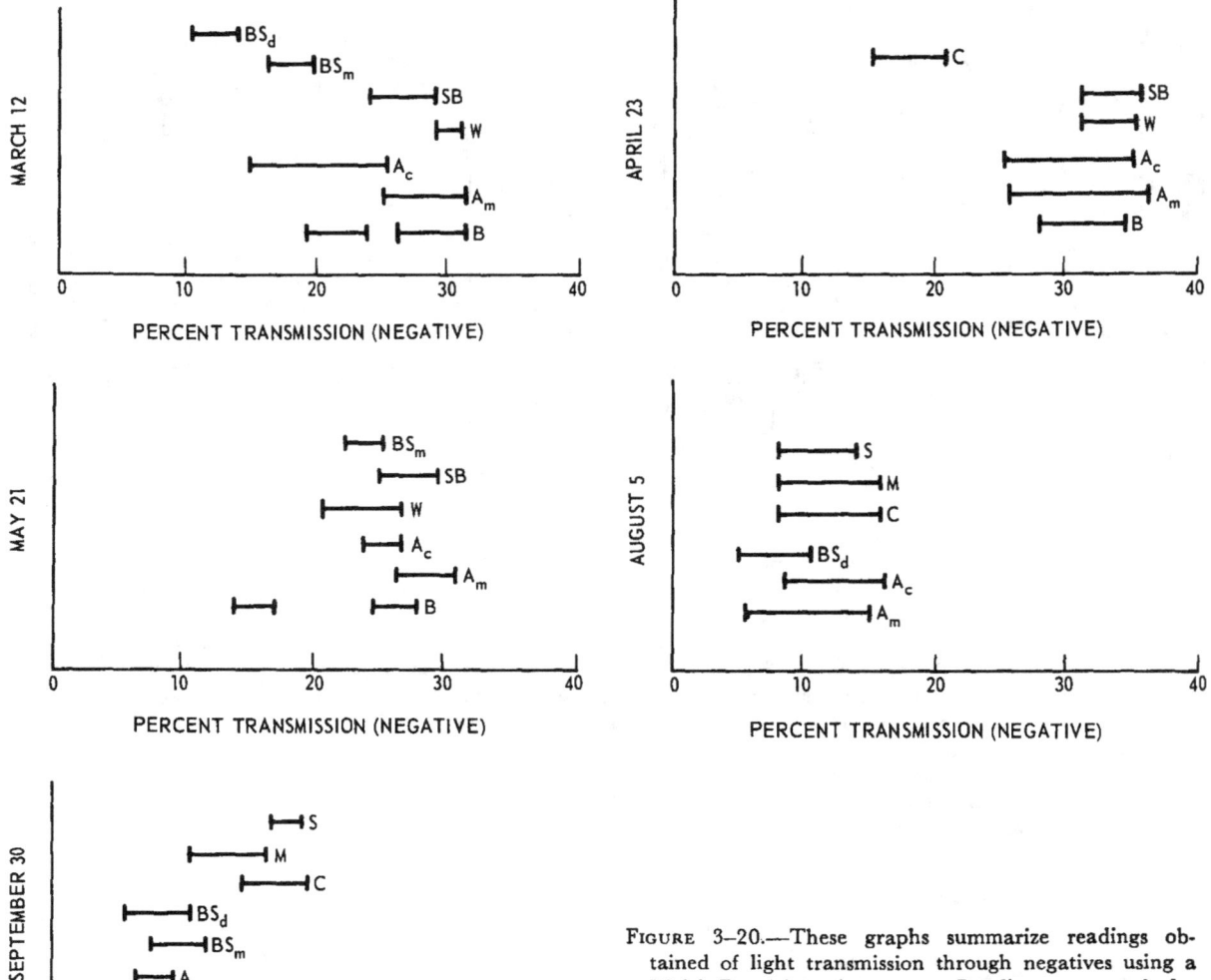

FIGURE 3–20.—These graphs summarize readings obtained of light transmission through negatives using a Welch Densichron instrument. Readings were made for several representative fields of each crop type in the 16-sq-mi Mesa test site on negatives made from Nikon 35-mm Pan-25A transparencies obtained on the dates indicated. (See the legend for figs. 3–2 to 3–16.)

tone on the Pan-25A print for May 21 (fig. 3–17), which is unique for that crop. Transmission values for barley (see the May 21 graph in fig. 3–20) are unique and do not overlap with values for other crops. As a result, interpreters were able to identify 91 percent of the barley fields correctly on the photograph for May 21, with only 3 percent commission errors.

In both of the above cases, interpretation results increased for the feature that was to be identified when there was a notable difference in the trans-

mission values for the crop in question as compared to all other crop types.

Data of the type just presented suggest that there are only a few times of year when particular crops can be successfully discriminated from all others on the basis of photographic tone. Furthermore, data of this type obtained for different film-filter combinations and for different dates could be useful in predicting the success expected from making either multidate or multiband color composites. Finally, these Densichron measurements lend credence to the observation that, for most crops on any given date, the within-crop variability of tone is greater than the between-crop variability. This is the problem that often limits one's success in identifying crops on one photo date, but is less restricting when sequential photos are available for evaluation of changing patterns of crop development.

Multiband and Multidate Color Composites

The potential for increasing the interpretability of space and aerial photographs for agricultural crop types through preparation of color composite images was investigated with photography of the Mesa test site. (The techniques for preparing images of this type were described in detail in ch. 2.) Multidate enhancements have been prepared that take advantage of the temporal signature of different crop types in the same manner as multiband enhancements are prepared to use spectral differences in tone signature. An optical color-combining system, the FRSL optical color combiner, and two electronic systems, the Philco-Ford image-tone enhancement system and the University of Kansas electronic processing system (image discrimination, enhancement, combination, and sampling (IDECS)) were used in the illustrations of the technique reported here.

Enhancements From the Optical Color Combiner

Color combinations were prepared using the FRSL optical color combiner with the intent to improve the identification of individual crop types. A comparison was made between the various kinds of enhancements prepared using multiband images from Apollo 9 and high-altitude photography (Color Plate 6a). Both sets of images were processed using similar projection filters and were prepared with the objective of enhancing overall interpretability of the image. Differences between the two enhancements are primarily due to differences in resolution of the input images.

In addition, enhancements were prepared to enhance individual crop types, using as inputs the same Pan-25A high-altitude-aircraft images that were selected as best for the discrimination of each crop type, based on the crop calendar. Thus, enhancements for cotton were prepared using March 12 and September 30 photographs; enhancements

for alfalfa and sugar beets were made using May 21 and September 30 photographs; and enhancements for barley were made using March 12 and May 21 photographs.

Color Plate 7 contains some of the examples referred to above. The enhanced images represent only three of many possible combinations. Conclusions reached regarding these three enhancements are only indicative of the possibility for improvement of the identification of crops by the enhancement technique. Filter choices in the FRSL optical color combiner are limited, and no adjustment of brightness for each projector is possible; more interpretable composites might be prepared from the same inputs if better adjustments were built into the system. Changes in the FRSL system used are being made. For each image, a map is included that shows the location of all fields of the crop for which the enhanced image was prepared. A photointerpreter was asked to study each image and determine how many fields of the crop type in question had been successfully color coded for identification by the enhancement process. The results of this comparison are shown in table 3–4. Percent correct identification and percent commission errors are tabulated for (1) the enhanced images, (2) the Infrared Ektachrome images that were judged optimum for distinguishing each crop type, and (3) Infrared Ektachrome images for March 12, April 23, and May 21, which were interpreted in concert. Results of interpreting the enhancements can be compared with interpretation of the same input images (Infrared Ektachrome photographs from the same dates as used to prepare the enhanced images), the results of which also appear in table 3–16. Since interpretation results from the images judged optimum were sometimes lower than for the concurrent interpretation of March 12, April 23, and May 21 images, results from that concurrent interpretation are also included. In this way we can compare interpretation results from enhanced images with the best results from interpretation of the Infrared Ektachrome photographs. The reader is invited to study each image in Color Plate 7, using the corresponding map to aid in determining how well each crop type is enhanced on the appropriate image. The following conclusions can be made regarding this comparison:

TABLE 3–4.—*Percent Correct Identification and Commission Errors for Enhanced Images*

Image identification	Percent of identification, by crop			
	Barley	Cotton	Alfalfa	Sugar beets
Percent correct:				
1. Enhanced images	86	81	92	62
2. Optimum multidate photos (Infrared Ektachrome)	66	84	70	32
3. Concurrent 3-date photos (Infrared Ektachrome)	82	(a)	83	35
Percent commission errors:				
1. Enhanced images	3	40	42	33
2. Optimum multidate photos (Infrared Ektachrome)	15	10	16	55
3. Concurrent 3-date photos (Infrared Ektachrome)	4	(a)	21	72

a No data available.

(1) *Barley.*—Barley fields have a unique blue color on the composite image prepared (March 12, 90 filter; May 21, 75 filter) to enhance that crop. Percent correct identification of barley on the enhanced image (86 percent) is similar to results obtained from concurrent interpretation of multidate photographs (82 percent). Also, only 3 percent commission errors were made (i.e., in such instances another crop was color coded in the same manner as barley).

(2) *Cotton.*—Cotton fields appear on this enhanced image (March 12, 75 filter; September 30, 25 filter) as bright blue or mottled blue fields. Percent correct identification for the enhancement (81 percent) is similar to that from multidate interpretation (84 percent), although commission errors are much higher (40 percent) for the enhanced image.

(3) *Alfalfa and sugar beets.*—Improvements in identification of these crops were also noted from a study of the enhancement made for these crops (May 21, 35 filter; September 30, 90 filter). Alfalfa appears very dark purple (not light purple or lavender) and sugar beets are bright yellow. Commission errors were higher for alfalfa and lower for sugar beets using the enhanced image in contrast to the multidate interpretation.

In conclusion, it can be said that the preparation of color composite images using multidate photography in an optical combiner system shows potential for increasing the accuracy with which crop types can be distinguished. The preceding discussion contains examples of cases in which interpretation results from enhanced images were comparable to, if not better than, results from the interpretation of multidate photographs. However, no statistical evaluation is attempted because of the differences in the manner in which the interpretation tests were performed.

Enhancements From the Philco-Ford System

Enhancements were also made for improving crop identification using the Philco-Ford image-tone enhancement system (described in ch. 2). The object was to isolate individual crop types by giving them unique color signatures, using multidate black-and-white high-altitude aircraft images as input material. Color Plate 8 contains examples of the output of this system (images (*a*), (*b*), and (*c*)). Again, it must be noted that these are only three of many possible combinations, and the results from examining these images are only representative of the potential of the Philco-Ford system. The meanings of the color signatures of the three images are tabulated in table 3–5.

The data in table 3–5 indicate the percentage of fields in each crop type that were color coded with the colors indicated. For example, 49 percent of the bare soil fields have a bright-green color and 40 percent of the bare soil fields have a mottled-green color on image (*a*). Fields of no other crop type have this color. In all, 89 percent of the bare soil fields are thus color coded as some shade of green. The data indicate that some crop types, but not all, are easily distinguishable on particular images—bare soil on images (*a*) and (*b*), barley on image (*c*)—by a characteristic color that is almost exclusive for that crop. The fact that image (*a*) or (*b*) is best for identifying bare soil and that image (*c*) is best for identifying barley reinforces earlier

TABLE 3–5.—*Color-Coded Fields, Philco-Ford Image-Tone Enhancement System*

Image	Color	Percent of crop category			
		Barley	Alfalfa	Bare soil	Sugar beets
(a) March (Pan-25A)+April (Pan-25A)	Bright green			49	
	Mottled green			40	
Total				89	
(b) March (IR–89B)+April (IR–89B)	Yellow			83	
	Red	3	32		
	Brownish-green	49	6		
Total		52	38	83	
(c) March (Pan-25A)+May (IR–89B)	Yellow		·17	20	12
	Brown	90		6	
Total		90	17	26	12

NOTE.—See Color Plate 8 for the images described here.

statements that the choice of image dates is important for identifying specific crop types and that techniques for manipulating the enhancement permit an individual crop type to be clearly separated from all others.

The overall interpretability of enhancements made from the Philco-Ford system using Apollo 9 and high-altitude photography was also compared (Color Plate 6(b)). The input images were subjected to similar enhancement procedures so that the effect of differing resolution of the input images could be compared. The resulting images show similar color signatures for the same fields, suggesting that the difference between the enhancements is primarily due to resolution differences.

Enhancements From the IDECS System

The IDECS system at the University of Kansas was also used to prepare composite images, as described in chapter 2. However, no multidate material was used. Instead, multiband photographs (Pan-25A, Pan-58, and IR–89B) taken on March 12 only were used to make the composite images that appear in Color Plate 8, images (d) and (e). These images were prepared to enhance a single crop type from all others. Image (d) was prepared to enhance the barley fields (blue). Image (e) was prepared to enhance all bare soil fields

(pale blue). Comparison with the adjoining maps in that figure gives the reader some indication of how successful this technique is for separating one crop type from all others as a unique color. The low resolution of the IDECS images makes it difficult to evaluate them on a field-by-field basis. However, the flexibility of using the controls of this device to select optimum density levels for enhancing specific features gives it great potential for producing image enhancements of high interpretability.

The examples from three image-enhancement systems that have just been discussed indicate the relative resolution of each system and the ability of each device to differentiate one crop type from another. These examples also illustrate how multidate enhancements can be prepared in much the same manner as multiband enhancements, and they offer the reader a means by which he can evaluate the advantages and disadvantages of each of three enhancement systems.

Interpretation Tests

To determine quantitatively the kinds of images that are most useful for identifying crop types, an interpretation test was prepared. The test consisted basically of the following:

(1) Fifteen test images that included four film-

filter combinations from the Apollo 9 flight and from the concurrent high-altitude mission that was flown in March 1969, sequential high-altitude photographs taken in the months of April and May 1969, and various multiband and multidate color composites. In addition, Infrared Ektachrome images were tested for dates judged optimum for identifying certain types of crops.

(2) Twelve interpreters who looked at no more than five of the test images; each test image was interpreted by four interpreters.

(3) Seven crop categories. These consisted of barley (B), recently cut alfalfa (A_c), mature alfalfa (A_m), wheat (W), sugar beets (SB), moist bare soil (BS_m), and dry bare soil (BS_d). Cotton and sorghum categories were also added for the test, using photos judged optimum for identification.

From the total population of fields within the 16-sq-mi agricultural test site at Mesa, several from each crop type or category were selected as training samples to familiarize interpreters with field conditions within the test site. The training samples were selected, after an examination of all fields within a particular category, to represent the range of variability exhibited by each category. The location of each training sample is indicated on the agricultural crop training map for March 12, 1969, in figure 3–21. This crop training map was given to each photointerpreter so that he might first study the appearance of each crop category on the various kinds of imagery used in the test. Once the photointerpreter had become familiar with the appearance of training samples, he was asked to identify the remaining unannotated fields within the test site. Table 3–6 shows the test images of the Mesa test site that were examined during the course of the photointerpretation test.

To minimize familiarity with the test site, the interpretation test was designed so that no interpreter would look at more than five of the 15 test images. Of the five images examined by an interpreter, no more than two were selected either from the black-and-white images, the Infrared Ektachrome images, or the color composite images. (An example of five images selected for an interpreter might include test images 1, 5, 6, 10, and 13.) The 12 interpreters (all having had previous experience with similar interpretation tests) were assigned

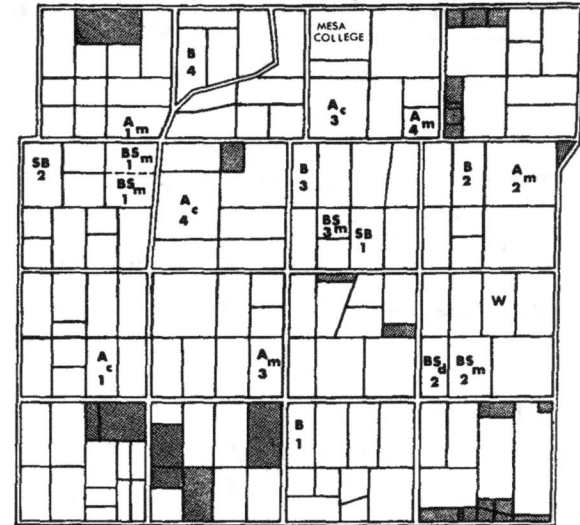

FIGURE 3–21.—Agricultural crop training map. In the design of the interpretation test to evaluate the agricultural resource, certain fields were selected from the population of known fields (shown in fig. 3–2) to be used as training samples. These training samples were selected as being representative of the range of variability that can be expected for a given crop type or condition at one date. An interpreter preparing to take the agricultural crop test first studied the tone or color and texture of the training samples. Once he had familiarized himself with the photoimage characteristics for each crop type, he attempted to identify the remaining fields not labeled on the training map. Fifteen test images of the Mesa agricultural test site (including black-and-white and Infrared Ektachrome space and aerial photographs and color composites made from the black-and-white photos) were examined in the interpretation test. The results of these tests are found in tables 3–7 to 3–16.

at random to interpret the various combinations of test images (such as those listed in the example above). By doing so, each test image was then examined by four different interpreters. The interpretation results (expressed as the cumulative number of fields seen by all four interpreters for each test image) can be found in tables 3–7 through 3–15. These tables are so prepared as to facilitate comparison of results from image types that are of interest to this study. Table 3–16 is a summary table that expresses in percentages the total correct identifications for each image type and the correct identifications for each category within each image type.

TABLE 3–6.—*Test Images Used in Photointerpretation Test, Mesa Test Site, 1969*

Test image no.	Imagery type	Date	Where illustrated
1	Pan–25A, Apollo 9	Mar. 12	Color Plate 3
2	IR–89B, Apollo 9	Mar. 12	Color Plate 3
3	Pan–25A, high altitude	Mar. 12	fig. 3–17
4	Pan–58, high altitude	Mar. 12	fig. 3–17
5	IR–89B, high altitude	Mar. 12	fig. 3–17
6	Infrared Ektachrome, Apollo 9	Mar. 12	Color Plate 3
7	Infrared Ektachrome, high altitude	Mar. 12	Color Plate 3, 5
8	Infrared Ektachrome, high altitude	Apr. 23	Color Plate 5
9	Infrared Ektachrome, high altitude	May 21	Color Plate 5
10	FRSL color composite, high altitude	Mar. 12	Color Plate 6(a)
11	FRSL color composite, Apollo 9	Mar. 12	Color Plate 6(a)
12	Philco-Ford color composite, high altitude	Mar. 12	Color Plate 6(b)
13	Philco-Ford color composite, Apollo 9	Mar. 12	Color Plate 6(b)
14	Multidate FRSL color composite (Pan–25A, high altitude)	March, May	Color Plate 7
15	Infrared Ektachrome, high altitude; images 7, 8, and 9 interpreted concurrently	March, April, May	Color Plate 5

Interpretation Results

Within each of the tables that express interpretation results (tables 3–7 through 3–15) the individual test image interpretations are presented in box form. Consequently the reader can compare the results of identifying separate crop categories by four interpreters (data along the rows) with the actual ground truth (data down the columns).

First, with reference to image 1 (table 3–7), let us consider the example in which the field actually was barley and note how it was interpreted as being by one or more photointerpreters. This is done by looking vertically down the first column, marked B for barley, in table 3–7. Note that 45 barley fields were correctly identified as barley (B) ; 45 barley fields were incorrectly identified as mature alfalfa (A_m) ; 13 barley fields were incorrectly identified as recently cut alfalfa (A_c) ; 17 barley fields were incorrectly identified as sugar beets (SB) ; 5 barley fields were incorrectly identified as wheat (W) ; 5 barley fields were incorrectly called moist bare soil (BS_m), and, finally, 1 barley field was incorrectly called dry bare soil (BS_d). Hence, of the 131 barley fields, 45 were identified correctly, yielding a percent-correct rating for barley of 34 percent. (See table 3–16.) (The "percent correct identification" for barley was computed by dividing the number of barley fields identified correctly by the total number of barley fields, and multiplying the quotient by 100; 45/131 × 100 = 34 percent.) The remaining 86 barley fields (sum of 45, 13, 17, 5, 5, and 1) are incorrect identifications called *omission errors*.

Next, with reference to image 1 (table 3–7), let us consider what is the true identity of each field that one or more of the interpreters called barley (B). This is done by looking *horizontally* along the first row, marked (B) for barley. Forty-five barley fields were correctly identified as barley. However, 13 fields of mature alfalfa (A_m) were incorrectly identified as barley; 10 recently cut alfalfa (A_c) fields were incorrectly identified as barley; and finally, the interpreters twice incorrectly identified sugar beets (SB) and wheat (W) as barley. Hence, the interpreters identified a total of 72 fields of barley, of which 45 were correct. The remaining 27 fields (sum of 13, 10, 2, and 2) which were incorrectly called barley are called *commission errors*. The cumulative number of actual barley fields in the Mesa test site is 131 (33 actual fields × 4 interpreters = 132 total barley fields minus one oversight).

In the overall summary shown in table 3–16, a commission error of 37 percent is recorded for barley as interpreted from image 1. This value was computed by dividing the number of commission errors made by the photointerpreters for barley (27, as stated in the preceding paragraph) by the number of barley fields seen by the interpreters and multiplying the quotient by 100; thus 27/72 × 100 = 37 percent commission error. When the

TABLE 3–7.—*Test Results Comparing the Interpretability of Agricultural Crop Types on Two Black-and-White Film-Filter Combinations and Infrared Ektachrome Space Photos*

Image 1: Pan-25A, Apollo 9, Mar. 12, 1969

Photointerpreter's results	Ground truth							Total fields	Number of errors
	B	Am	Ae	SB	W	BSm	BSd		
B	*45*	13	10	2	2			72	27
Am	45	*34*	11	15	1		1	107	73
Ae	13	5	*37*	3	1	12	8	79	42
SB	17	9		*4*				30	26
W	5	1			*0*			6	6
BSm	5		27			*33*	22	87	54
BSd	1	2	7			7	*37*	54	17
Total fields	131	64	92	24	4	52	68	435	
Incorrect	86	30	55	20	4	19	31		245

Total percentage correct identification: 43

Image 6: Infrared Ektachrome, Apollo 9, Mar. 12, 1969

Photointerpreter's results	Ground truth							Total fields	Number of errors
	B	Am	Ae	SB	W	BSm	BSd		
B	*66*	21	1	4				92	26
Am	31	*33*	2	3	2			71	38
Ae	22	5	*84*	10			3	124	40
SB	5	3	2	*6*				16	10
W	7	2		1	*2*			12	10
BSm						*43*	14	57	14
BSd		1		3		9	*51*	64	13
Total fields	132	64	92	24	4	52	68	436	
Incorrect	66	31	8	18	2	9	17		151

Total percentage correct identification: 65

Image 2: IR-89B, Apollo 9, Mar. 12, 1969

Photointerpreter's results	Ground truth							Total fields	Number of errors
	B	Am	Ae	SB	W	BSm	BSd		
B	*35*	13		2	4			54	19
Am	46	*31*	11	1				89	58
Ae	25	14	*54*	7		10	3	113	59
SB	18	3	7	*10*				38	28
W	6	2	2	1	*0*			11	11
BSm	1			2		*22*	9	34	12
BSd	1	1	17			20	*56*	95	39
Total fields	132	64	91	23	4	52	68	434	
Incorrect	97	33	37	13	4	30	12		226

Total percentage correct identification: 47

NOTE.—Numbers in the bodies of tables 3–7 through 3–15 indicate the cumulative number of fields identified by four interpreters. Numbers in italic indicate the number of fields identified correctly. To illustrate the use of these tables, consider image 1 of this table. The numbers in the upper left corner have the following significance: 45 of the known (from ground data) barley fields (B) were correctly identified as barley by the interpreters; 34 mature alfalfa fields (Am) were correctly identified as mature alfalfa. These 2 numbers are italicized. However, 45 of the known barley fields (first column, second row) were incorrectly called mature alfalfa, and 13 of the mature alfalfa fields were incorrectly called barley by the interpreters. B = barley, Am = mature alfalfa, Ae = recently cut alfalfa, SB = sugar beets, W = wheat, BSm = moist bare soil, BSd = dry bare soil.

percent commission error for a particular category is relatively high, it means that other crop categories are frequently misidentified as the category in question.

An *overall rating of interpretability* for each test image is expressed as the total percentage of correct identifications for all categories. This value is computed by summing the correct identifications for all categories (values in italic that form a diagonal row, as in table 3–7 for example), dividing this by the total number of fields and multiplying the result by 100. Thus an overall interpretability rating for image 1 is seen to be 43 percent as read in the "percent correct" column of table 3–16 and in the note to image 1 in table 3–7.

Table 3–7 also presents interpretation results that will allow comparisons to be made between the Apollo 9 black-and-white (Pan-25A and IR-89B) and Infrared Ektachrome photographs (test images 1, 2, and 6), for identifying crop categories. For overall interpretability the Infrared Ektachrome photo (65 percent correct identification) appears to be better than the two black-and-white photographs, Pan-25A and IR–89B (43 and 47 percent, respectively), which themselves do not appear to be very different. The increase in interpretability of the Infrared Ektachrome photograph is attributable to the increased accuracy of identifying recently cut alfalfa and bare soil categories. Generally, however, the relatively low percentage of correct identifications for barley, alfalfa, sugar beets, and wheat suggest that they cannot be accurately discriminated on any of the space photos taken in early March. On the Infrared Ektachrome space photo it nearly always was possible for interpreters to discriminate bare soil from fields containing a growing cover crop.

Table 3–8 presents interpretation results that

TABLE 3–8. *Test Results Comparing the Interpretabili'y of Agricultural Crop Types on 2 Black-and-White Space Photographs and on 2 Color Composites (FRSL and Philco-Ford) Made From the Black-and-White Images*

Image 1: Pan-25A, Apollo 9, Mar. 12, 1969

Photointerpreter's results	Ground truth							Total fields	Number of errors
	B	Am	Ac	SB	W	BSm	BSd		
B	45	13	10	2	2			72	27
Am	45	34	11	15	1		1	107	73
Ac	13	5	37	3	1	12	8	79	42
SB	17	9		4				30	26
W	5	1			0			6	6
BSm	5		27			33	22	87	54
BSd	1	2	7			7	37	54	17
Total fields	131	64	92	24	4	52	68	435	
Incorrect	86	30	55	20	4	19	31		245

Total percentage correct identification: 43

Image 11: FRSL Color Composite, Apollo 9, Mar. 12, 1969

Photointerpreter's results	Ground truth							Total fields	Number of errors
	B	Am	Ac	SB	W	BSm	BSd		
B	52	10		3	3			68	16
Am	30	33	10	5	1		2	81	48
Ac	19	10	57	11			4	101	44
SB	27	8	5	5				45	40
W	4	1			0			5	5
BSm			2			30	19	51	21
BSd			17			22	42	81	39
Total fields	132	62	91	24	4	52	67	432	
Incorrect	80	29	34	19	4	22	25		213

Total percentage correct identification: 50

Image 2: IR–89B, Apollo 9, Mar. 12, 1969

Photointerpreter's results	Ground truth							Total fields	Number of errors
	B	Am	Ac	SB	W	BSm	BSd		
B	35	13		2	4			54	19
Am	46	31	11	1				89	58
Ac	25	14	54	7		10	3	113	59
SB	18	3	7	10				38	28
W	6	2	2	1	0			11	11
BSm	1			2		22	9	34	12
BSd	1	1	17			20	56	95	39
Total fields	132	64	91	23	4	52	68	434	
Incorrect	97	33	37	13	4	30	12		226

Total percentage correct identification: 47

Image 13: Philco-Ford Color Composite, Apollo 9, Mar. 12, 1969

Photointerpreter's results	Ground truth							Total fields	Number of errors
	B	Am	Ac	SB	W	BSm	BSd		
B	32	10	7	2		9	15	75	43
Am	21	25	5	1	1	3		56	31
Ac	28	13	26	8	1	14	10	100	74
SB	41	13	17	8		4	5	88	80
W	3	1			2			6	4
BSm	6	1	24	3		72		46	34
BSd	1	1	13	2		10	38	65	27
Total fields	132	64	92	24	4	52	68	436	
Incorrect	100	39	66	16	2	40	30		293

Total percentage correct identification: 32
NOTE.—See table 3–7 for explanation of data.

permit comparison of the two black-and-white space photos (Pan-25A and IR–89B) and two color composites (FRSL and Philco-Ford) made from them. Noting first the overall interpretability ratings, one might conclude that there is a slight improvement in interpretability on an FRSL color composite versus the two black-and-white space photos; however, these differences are probably not significant. The low rating for the Philco-Ford color composite suggests that this particular composite, which represents only one of an infinite number of possible composites, did not in fact enhance the interpretability of the various crop categories. Again, since the overall interpretability ratings for these four images are relatively low (i.e., 50 percent and below), it is concluded that accurate category discrimination is not well performed from these space images obtained in March 1969.

Table 3–9 presents the interpretation results comparing an Apollo 9 Infrared Ektachrome photograph with two color composites made from the black-and-white Apollo 9 photographs. The Infrared Ektachrome photograph appears to be more interpretable than the color composites, judging from the total percent correct identifications for all categories (65 percent versus 50 and 32 percent). Again, the increase can be attributable to the fact that recently cut alfalfa (A_c) and bare soil (BS) are readily identified on the Infrared Ektachrome photograph.

Table 3–10 compares the interpretation results of the four film-filter combinations (Pan-25A, Pan-58, IR–89B, and Infrared Ektachrome) procured by a high-altitude aircraft at the same time as the Apollo 9 mission. It is of interest to note that the overall interpretability rating of these film-filter combina-

TABLE 3–9.—*Test Results Comparing the Interpretability of Agricultural Crop Types on an Infrared Ektachrome Space Photograph, an FRSL Color Composite, and a Philco-Ford Color Composite*

Image 11: FRSL Color Composite, Apollo 9, Mar. 12, 1969

Photointerpreter's results	Ground truth							Total fields	Number of errors
	B	Am	Ac	SB	W	BSm	BSd		
B	52	10	-----	3	3	----------		68	16
Am	30	33	10	5	1	-----	2	81	48
Ac	19	10	57	11	----------		4	101	44
SB	27	8	5	5	----------------			45	40
W	4	1	----------		0	----------		5	5
BSm			2	----------		30	19	51	21
BSd			17	----------		22	42	81	39
Total fields	132	62	91	24	4	52	67	432	------
Incorrect	80	29	34	19	4	22	25	------	213

Total percentage correct identification: 50

Image 13: Philco-Ford Color Composite, Apollo 9, Mar. 12, 1969

Photointerpreter's results	Ground truth							Total fields	Number of errors
	B	Am	Ac	SB	W	BSm	BSd		
B	32	10	7	2	-----	9	15	75	43
Am	21	25	5	1	1	3	-----	56	31
Ac	28	13	26	8	1	14	10	100	74
SB	41	13	17	8	-----	4	5	88	80
W	3	1	----------		2	----------		6	4
BSm	6	1	24	3	-----	12		46	34
BSd	1	1	13	2	-----	10	38	65	27
Total fields	132	64	92	24	4	52	68	436	------
Incorrect	100	39	66	16	2	40	30	------	293

Total percentage correct identification: 32

Image 6: Infrared Ektachrome, Apollo 9, Mar. 12, 1969

Photointerpreter's results	Ground truth							Total fields	Number of errors
	B	Am	Ac	SB	W	BSm	BSd		
B	66	21	1	4	----------------			92	26
Am	31	33	2	3	2	----------		71	38
Ac	22	5	84	10	----------		3	124	40
SB	5	3	2	6	----------------			16	10
W	7	2	-----	1	2	----------		12	10
BSm						43	14	57	14
BSd		1	-----	3	----------		9	51	64	13
Total fields	132	64	92	24	4	52	68	436	------
Incorrect	66	31	8	18	2	9	17	------	151

Total percentage correct identification: 65
NOTE.—See table 3–7 for explanation of data.

tions is essentially the same whether the photographs were obtained from an aircraft or a spacecraft (compare results of table 3–7 with 3–10). The Infrared Ektachrome photograph is more interpretable than any of the black-and-white photographs when the images are studied individually. The overall interpretability of the three types of black-and-white aerial photography is about the same (47 percent versus 41 percent versus 45 percent for Pan-25A, Pan-58, and IR–89B). The infrared-sensitive films (IR–89B and Infrared Ektachrome) are more useful for discriminating bare soil from cover crops, but all the film-filter combinations are rather unreliable for separating the various cover crops. When the photos are studied individually, the Infrared Ektachrome photo is best for identifying recently cut alfalfa fields and bare soil.

Table 3–11 compares the interpretation results of three black-and-white aerial photographs with a color composite made by the FRSL optical combiner. It is apparent from the overall interpretability ratings of these four images that the color composite is more interpretable than the individual black-and-white bands.

Table 3–12 provides an opportunity to compare the test results of an Infrared Ektachrome high-altitude aerial photograph with three color composites; two of them made using black-and-white images on different enhancement systems (FRSL and Philo-Ford) and the other made using two Pan-25A images taken in March and in May. A comparison of the Infrared Ektachrome photograph with the FRSL color composite shows that there is little difference in overall interpretability (64 percent versus 58 percent). The Infrared Ektachrome photograph, however, is slightly better for discriminating bare soil from the cover crops. The overall interpretability rating for the Philco-Ford color composite is lower than that for either of the previously discussed images. This may be due in part to lower resolution of the image, or perhaps to using a less-than-optimum color composite for discriminating crop categories.

The multidate color composite is of particular interest because an obvious improvement in interpretability was achieved on this image (76 percent correct identification) compared to the other three images, by virtue of the addition of the time di-

TABLE 3-10.—*Test Results Comparing the Interpretability of Agricultural Crop Types on 3 Black-and-White Infrared Ektachrome High-Altitude Aerial Photographs*

Image 3: Pan-25A, High Altitude, Mar. 12, 1969

Photointerpreter's results	Ground truth							Total fields	Number of errors
	B	Am	Ac	SB	W	BSm	BSd		
B	44	17	4	2	4			71	27
Am	43	35	9	1				88	53
Ac	10	1	37	7		10	4	69	32
SB	20	11	6	10				47	37
W	5			1	0			6	6
BSm	5		18	2		22	5	52	30
BSd	3		18			20	59	100	41
Total fields	130	64	92	23	4	52	68	433	
Incorrect	86	29	55	13	4	30	9		226

Total percentage correct identification: 47

Image 5: IR-89B, High Altitude, Mar. 12, 1969

Photointerpreter's results	Ground truth							Total fields	Number of errors
	B	Am	Ac	SB	W	BSm	BSd		
B	43	20	2	4	4		1	74	31
Am	38	34	16	7		1	1	97	63
Ac	13	1	38	4		3	10	69	31
SB	30	3	17	6				56	50
W	7	6	2	3	0			18	18
BSm						38	17	55	17
BSd	1		17			8	39	65	26
Total fields	132	64	92	24	4	50	68	434	
Incorrect	89	30	54	18	4	12	29		236

Total percentage correct identification: 45

Image 4: Pan-58, High Altitude, Mar. 12, 1969

Photointerpreter's results	Ground truth							Total fields	Number of errors
	B	Am	Ac	SB	W	BSm	BSd		
B	39	14	16	1	3	1		74	35
Am	26	28	4	4		4		66	38
Ac	37	11	52	19		20	11	150	98
SB	14	4	1	0		6	3	28	28
W	5	1	1		1			8	7
BSm	7	6	7			11	3	34	23
BSd	4		10			10	51	75	24
Total fields	132	64	91	24	4	52	68	435	
Incorrect	93	36	39	24	3	41	17		253

Total percentage correct identification: 41

Image 7: Infrared Ektachrome, High Altitude, Mar. 12, 1969

Photointerpreter's results	Ground truth							Total fields	Number of errors
	B	Am	Ac	SB	W	BSm	BSd		
B	43	17	1		4			65	22
Am	38	36	6	2				82	46
Ac	24	3	72	10				109	37
SB	24	5	2	11				42	31
W	2	3		1	0			6	6
BSm						39		39	0
BSd	1		9			13	67	90	23
Total fields	132	64	90	24	4	52	67	433	
Incorrect	89	28	18	13	4	13	0		152

Total percentage correct identification: 64
NOTE.—See table 3-7 for explanation of data.

mension. The multidate color composite (Color Plate 7) is composed of a Pan-25A image taken in March and a Pan-25A image taken in May. On the May image, most of the barley fields exhibit a distinctive signature that enables an interpreter to improve his accuracy for identifying barley. In turn, the interpreter increased the correct identifications for alfalfa; thus, the overall rating was increased. When such an image, e.g., May, Pan-25A, is among those used in making a multidate color composite, the effect is to increase the overall interpretability of that image when compared to images made at some other date (e.g., March) when discrimination among crops is difficult.

Table 3-13 contrasts the results of two black-and-white photos (Pan-25A and IR-89B) obtained from Apollo 9 with those same bands obtained from the high-altitude aircraft. The interesting conclusion based upon their overall interpretability ratings is that, for discriminating crop categories, there is little difference between the black-and-white space photographs and the black-and-white aerial photos obtained in March. The overall interpretability ratings are remarkably similar (43, 47, 47, and 45 percent).

Table 3-14 compares the interpretation results of the space and aerial photographs that were obtained with Infrared Ektachrome film on March 12, 1969. Notice that here, too, the overall interpretability ratings for these two images, expressed as the total percent correct identifications for all categories, are nearly the same (65 percent versus 64 percent). This suggests that Infrared Ektachrome space photographs are as interpretable as aerial photographs for the identification of crop categories in March. For both image types the bare soil category (see table 3-16) is almost always

TABLE 3–11.—*Test Results Comparing the Interpretability of Crop Types on 3 Black-and-White Film-Filter Combinations (High-Altitude Aircraft) and a Color Composite (FRSL Optical Combiner) Made From the Same Film-Filter Combinations*

Image 3: Pan-25A, High Altitude, Mar. 12, 1969

Photointerpreter's results	Ground truth							Total fields	Number of errors
	B	Am	Ac	SB	W	BSm	BSd		
B	44	17	4	2	4			71	27
Am	43	35	9	1				88	53
Ac	10	1	37	7		10	4	69	32
SB	20	11	6	10				47	37
W	5			1	0			6	6
BSm	5		18	2		22	5	52	30
BSd	3		18			20	59	100	41
Total fields	130	64	92	23	4	52	68	433	
Incorrect	86	29	55	13	4	30	9		226

Total percentage correct identification: 47

Image 5: IR–89B, High Altitude, Mar. 12, 1969

Photointerpreter's results	Ground truth							Total fields	Number of errors
	B	Am	Ac	SB	W	BSm	BSd		
B	43	20	2	4	4		1	74	31
Am	38	34	16	7		1	1	97	63
Ac	13	1	38	4		3	10	69	31
SB	30	3	17	6				56	50
W	7	6	2	3	0			18	18
BSm						38	17	55	17
BSd	1		17			8	39	65	26
Total fields	132	64	92	24	4	50	68	434	
Incorrect	89	30	54	18	4	12	29		236

Total percentage correct identification: 45

Image 4: Pan-58, High Altitude, Mar. 12, 1969

Photointerpreter's results	Ground truth							Total fields	Number of errors
	B	Am	Ac	SB	W	BSm	BSd		
B	39	14	16	1	3	1		74	35
Am	26	28	4	4		4		66	38
Ac	37	11	52	19		20	11	150	98
SB	14	4	1	0		6	3	28	28
W	5	1	1		1			8	7
BSm	7	6	7			11	3	34	23
BSd	4		10			10	51	75	24
Total fields	132	64	91	24	4	52	68	435	
Incorrect	93	36	39	24	3	41	17		253

Total percentage correct identification: 41

Image 10: FRSL Color Composite, High Altitude, Mar. 12, 1969

Photointerpreter's results	Ground truth							Total fields	Number of errors
	B	Am	Ac	SB	W	BSm	BSd		
B	47	8		1	2			58	11
Am	57	51	15	8	2			133	82
Ac	23	3	57	8				91	34
SB	4		5	5				14	9
W	1	2		1	0			4	4
BSm			11			38	11	60	22
BSd			4			13	57	74	17
Total fields	132	64	92	23	4	51	68	434	
Incorrect	85	13	35	18	4	13	11		179

Total percentage correct identification: 58
NOTE.—See table 3–7 for explanation of data.

discriminated from cover-crop categories, but cover crops are not accurately differentiated.

Table 3–15 shows the interpretation results from three Infrared Ektachrome aerial photographs, taken in March, April, and May, analyzed both separately and concurrently. Of the three images taken on different dates, the image obtained in May gave the highest overall interpretability rating. This is ascribed to the increased ability to identify barley fields, which have a unique color signature at that date. Nevertheless, there is slightly more misidentification of bare soil with recently cut alfalfa fields than exists on the March or April images. Also, as has been indicated earlier, the increased ability to identify one crop, in this case barley, also increases the ability to separate and identify other crops, e.g., alfalfa, which at an earlier date may have been confused with barley.

When all three Infrared Ektachrome images from the three dates were examined concurrently by four interpreters, the highest overall correct identifications for all crop categories was obtained. For this particular test the interpreters were asked to identify the crop category that existed in March, based upon the sequential changes that had taken place during the 2-month interval between the March and May photographs. Not only were the percentages of correct identifications high for barley, alfalfa, and bare soil, but also the commission errors were low. Hence, the conclusion based upon the images compared in table 3–15 is that when sequential images are interpreted in concert, acceptably high correct identifications of crop categories can be made on very-high-altitude aerial photos. From a comparison of the results obtained by interpreting space with aerial photos, as summarized in tables 3–13 and 3–14, it is again concluded that interpretation of sequential images

TABLE 3-12.—*Test Results Comparing the Interpretability of Agricultural Crop Types on an Infrared Ektachrome High-Altitude Aerial Photograph, an FRSL Color Composite, a Philco-Ford Color Composite, and a Color Composite Made From Pan-25 Photos Taken in March and May*

Image 7: Infrared Ektachrome, High Altitude, Mar. 12, 1969

Photointerpreter's results	Ground truth							Total fields	Number of errors
	B	Am	Ae	SB	W	BSm	BSd		
B	43	17	1	-----	4	-----	-----	65	22
Am	38	36	6	2	-----	-----	-----	82	46
Ae	24	3	72	10	-----	-----	-----	109	37
SB	24	5	2	11	-----	-----	-----	42	31
W	2	3	-----	1	0	-----	-----	6	6
BSm	-----	-----	-----	-----	-----	39	-----	39	0
BSd	1	-----	9	-----	-----	13	67	90	23
Total fields	132	64	90	24	4	52	67	433	
Incorrect	89	28	18	13	4	13	0		152

Total percentage correct identification: 64

Image 10: Color Composite FRSL Optical Combiner, High Altitude, Mar. 12, 1969

Photointerpreter's results	Ground truth							Total fields	Number of errors
	B	Am	Ae	SB	W	BSm	BSd		
B	47	8	-----	1	2	-----	-----	58	11
Am	57	51	15	8	2	-----	-----	133	82
Ae	23	3	57	8	-----	-----	-----	91	34
SB	4	-----	5	5	-----	-----	-----	14	9
W	1	2	-----	1	0	-----	-----	4	4
BSm	-----	-----	-----	11	-----	38	11	60	22
BSd	-----	-----	-----	4	-----	13	57	74	17
Total fields	132	64	92	23	4	51	68	434	
Incorrect	85	13	35	18	4	13	11		179

Total percentage correct identification: 58

Image 14: Multidate Color Composite, FRSL Optical Combiner, High Altitude, Mar. 12 and May 21, 1969

Photointerpreter's results	Ground truth							Total fields	Number of errors
	B	Am	Ae	SB	W	BSm	BSd		
B	102	1	-----	2	-----	-----	-----	105	3
A	7	113	-----	22	3	14	-----	159	46
Ae	-----	-----	0	-----	-----	-----	-----		--
SB	1	3	-----	0	-----	-----	-----	4	4
W	3	1	-----	-----	0	1	-----	5	5
BS	5	34	-----	-----	1	112	-----	152	40
Total fields	118	152	-----	24	4	127	-----	425	
Incorrect	16	39	-----	24	4	15	-----		98

Total percentage correct identification: 76

Image 12: Color Composite, Philco-Ford, High Altitude, Mar. 12, 1969

Photointerpreter's results	Ground truth							Total fields	Number of errors
	B	Am	Ae	SB	W	BSm	BSd		
B	30	6	-----	1	1	-----	-----	38	8
Am	45	42	9	3	2	-----	-----	101	59
Ae	24	3	43	14	-----	-----	3	87	44
SB	16	10	1	2	-----	-----	-----	29	27
W	7	2	-----	2	7	-----	-----	12	11
BSm	6	-----	27	2	-----	36	19	90	54
BSd	1	1	12	-----	-----	16	46	76	30
Total fields	129	64	92	24	4	52	68	433	
Incorrect	99	22	49	22	3	16	22		233

Total percentage correct identification: 46
NOTE.—See table 3-7 for explanation of data.

from space could, indeed, provide an acceptably high accuracy of crop identifications.

Now that the results of the interpretation test have been presented in terms of comparisons among certain of the test images (those compared in tables 3-7 to 3-15), it is important for us to recognize that specific conclusions based upon test results cannot automatically be drawn for some of the test images. As an example, consider test image 14, called a "multidate color composite." This particular composite was made using two Pan-25A photographs: one taken in March and the other in May. It represents only one out of innumerable composites that could be made from the combinations of black-and-white bands (Pan-25A, Pan-58, and IR-89B), dates of photography (March, April, May, August, etc.), and enhancement filters that could be selected to make multidate enhancements. Hence, before concluding that the multidate enhancement (image 14) is more or less interpretable than some other image (film-filter combination or color composite), it should be remembered that we are dealing with a single combination, and any other combination may be less interpretable or more interpretable, depending upon the selection of the spectral band, date of the photo, and combination of enhancement filters. The same kind of consideration must be given when evaluating the interpretation results for the "multiband color composites." For these, the interpretability of the color composite likewise is a function of the overall interpretability of the individual bands used to

TABLE 3–13.—*Test Results Comparing the Interpretability of Agricultural Crop Types on 2 Black-and-White Space Photographs (Pan-25A, IR–89B) with 2 Black-and-White High-Altitude Aerial Photographs (Pan-25A, IR–89B)*

Image 1: Pan-25A, Apollo 9, Mar. 12, 1969

Photointerpreter's results	Ground truth							Total fields	Number of errors
	B	Am	Ae	SB	W	BSm	BSd		
B	45	13	10	2	2			72	27
Am	45	34	11	15	1		1	107	73
Ae	13	5	37	3	1	12	8	79	42
SB	17	9		4				30	26
W	5	1			0			6	6
BSm	5		27			33	22	87	54
BSd	1	2	7			7	37	54	17
Total fields	131	64	92	24	4	52	68	435	
Incorrect	86	30	55	20	4	19	31		245

Total percentage correct identification: 43

Image 3: Pan-25A, High Altitude, Mar. 12, 1969

Photointerpreter's results	Ground truth							Total fields	Number of errors
	B	Am	Ae	SB	W	BSm	BSd		
B	44	17	4	2	4			71	27
Am	43	35	9	1				88	53
Ae	10	1	37	7		10	4	69	32
SB	20	11	6	10				47	37
W	5			1	0			6	6
BSm	5		18	2		22	5	52	30
BSd	3		18			20	59	100	41
Total fields	130	64	92	23	4	52	68	433	
Incorrect	86	29	55	13	4	30	9		226

Total percentage correct identification: 47

Image 2: IR–89B, Apollo 9, Mar. 12, 1969

Photointerpreter's results	Ground truth							Total fields	Number of errors
	B	Am	Ae	SB	W	BSm	BSd		
B	35	13		2	4			54	19
Am	46	31	11	1				89	58
Ae	25	14	54	7		10	3	113	59
SB	18	3	7	10				38	28
W	6	2	2	1	0			11	11
BSm	1			2		22	9	34	12
BSd	1	1	17			20	56	95	39
Total fields	132	64	91	23	4	52	68	434	
Incorrect	97	33	37	13	4	30	12		226

Total percentage correct identification: 47

Image 5: IR–89B, High Altitude, Mar. 12, 1969

Photointerpreter's results	Ground truth							Total fields	Number of errors
	B	Am	Ae	SB	W	BSm	BSd		
B	43	20	2	4	4		1	74	31
Am	38	34	16	7		1	1	97	63
Ae	13	1	38	4		3	10	69	31
SB	30	3	17	6				56	50
W	7	6	2	3	0			18	18
BSm						38	17	55	17
BSd	1		17			8	39	65	26
Total fields	132	64	92	24	4	50	68	434	
Incorrect	89	30	54	18	4	12	29		236

Total percentage correct identification: 45
NOTE.—See table 3–7 for explanation of data.

TABLE 3–14.—*Test Results Comparing the Interpretability of Agricultural Crop Types on an Infrared Ektachrome Space Photograph with an Infrared Ektachrome High-Altitude Aerial Photograph*

Image 6: Infrared Ektachrome, Apollo 9, Mar. 12, 1969

Photointerpreter's results	Ground truth							Total fields	Number of errors
	B	Am	Ae	SB	W	BSm	BSd		
B	66	21	1	4				92	26
Am	31	33	2	3	2			71	38
Ae	22	5	84	10			3	124	40
SB	5	3	2	6				16	10
W	7	2		1	2			12	10
BSm						43	14	57	14
BSd	1		3			9	51	64	13
Total fields	132	64	92	24	4	52	68	436	
Incorrect	66	31	8	18	2	9	17		151

Total percentage correct identification: 65

Image 7: Infrared Ektachrome, High Altitude, Mar. 12, 1969

Photointerpreter's results	Ground truth							Total fields	Number of errors
	B	Am	Ae	SB	W	BSm	BSd		
B	43	17	1		4			65	22
Am	38	36	6	2				82	46
Ae	24	3	72	10				109	37
SB	24	5	2	11				42	31
W	2	3		1	0			6	6
BSm						39		39	0
BSd	1		9			13	67	90	23
Total fields	132	64	90	24	4	52	67	433	
Incorrect	89	28	18	13	4	13	0		152

Total percentage correct identification: 64
NOTE.—See table 3–7 for explanation of data.

TABLE 3–15.—*Test Results Comparing the Interpretability of Agricultural Crop Types on Infrared Ek-tachrome Aerial Photographs Taken in March, April, and May, With These Images Interpreted in Concert*

Image 7: Infrared Ektachrome, High Altitude, Mar. 12, 1969

Photointerpreter's results	Ground truth						Total fields	Number of errors
	B	Am	Ac	SB	W	BS		
B	*43*	17	1			4	65	22
Am	38	*36*	6	2			82	46
Ac	24	3	*72*	10			109	37
SB	24	5	2	*11*			42	31
W	2	3		1	*0*		6	6
BS	1		9			*119*	129	10
Total fields	132	64	90	24	4	119	433	
Incorrect	89	28	18	13	4	0		152

Total percentage correct identification: 64

Image 8: Infrared Ektachrome, High Altitude Apr. 23, 1969

Photointerpreter's results	Ground truth						Total fields	Number of erros
	B	Am	Ac	SB	W	BSm		
B	*68*	8	8	2		3	89	21
Am	12	*32*	31	11	3		89	57
Ac	4	20	*32*	3		1	60	28
SB	18	4	6	*8*	1	1	38	30
W	13	3			*0*		16	16
BS	4	1	8			*115*	128	13
Total fields	119	68	85	24	4	120	420	
Incorrect	51	36	53	16	4	5		165

Total percentage correct identification: 60

Image 9: Infrared Ektachrome, High Altitude, May 21, 1969

Photointerpreter's results	Ground truth						Total fields	Number of errors
	B	Am	Ac	SB	W	BSm		
B	*108*			3		4	115	7
Am		*50*	21	8			79	29
Ac	6	1	*46*	2	4	12	71	25
SB		9	17	*8*		4	38	30
W					*0*	4	4	4
BS	6		16	3		*102*	127	25
Total fields	120	60	100	24	4	126	434	
Incorrect	12	10	54	16	4	24		120

Total percentage correct identification: 72

Image 15: Concurrent Interpretation of Images 7, 8, and 9

Photointerpreter's results	Ground truth						Total fields	Number of errors
	B	Am	Ac	SB	W	BSm		
B	*107*	1	1	2			111	4
Am	13	*50*	6	7	3	1	80	30
Ac	3	4	*70*	6		1	84	14
SB	4	9	8	*8*			29	21
W	1				*7*		2	1
BS	2		8			*116*	126	10
Total fields	130	64	93	23	4	118	432	
Incorrect	23	14	23	15	3	2		80

Total percentage correct identification: 81

NOTE.—See table 3–7 for explanation of data.

TABLE 3–16.—*Percent-Correct (and Percent-Commission-Error) Identifications for Individual and Overall Crop Types in an Interpretation Test of the Mesa Test Site, 1969*

Test image no.	Imagery type	Date	Total percent correct	Barley		Mature alfalfa		Cut alfalfa		Alfalfa		Sugar beets	
				Cor.	Com.	Cor.	Com.	Cor.	Com.	Cor.	Com.	Cor.	Com.
1	Pan–25A, Apollo 9	Mar. 12	43	34	37	53	68	40	53	56	53	16	86
2	IR–89B, Apollo 9	Mar. 12	47	27	35	48	65	59	52	71	48	43	74
3	Pan–25A, high altitude	Mar. 12	47	34	38	55	60	40	46	53	47	43	79
4	Pan–58, high altitude	Mar. 12	41	30	47	44	58	57	65	61	50	0	100
5	IR–89B, high altitude	Mar. 12	45	33	42	53	65	41	45	57	46	25	89
6	Infrared Ektachrome, Apollo 9	Mar. 12	65	50	28	52	54	91	33	80	34	25	62
7	Infrared Ektachrome, high altitude	Mar. 12	64	33	34	56	56	80	34	76	38	46	74
8	Infrared Ektachrome, high altitude	Apr. 23	60	57	24	47	64	38	47	75	22	33	79
9	Infrared Ektachrome, high altitude	May 21	72	90	6	83	37	44	35	72	21	40	79

TABLE 3–16 (Cont.)—*Percent-Correct (and Percent-Commission-Error) Identifications for Individual and Overall Crop Types in an Interpretation Test of the Mesa Test Site, 1969*

Test image no.	Imagery type	Date	Total percent correct	Wheat		Bare soil, moist		Bare soil, dry		Bare soil		Cereal	
				Cor.	Com.	Cor.	Com.	Cor.	Com.	Cor.	Com.	Cor.	Com.
10	FRSL color composite, high altitude	Mar. 12	58	36	19	80	62	62	37	81	43	22	64
11	FRSL color composite, Apollo 9	Mar. 12	50	39	24	53	59	63	44	72	39	21	89
12	Philco-Ford color composite, high altitude	Mar. 12	46	23	21	66	58	47	51	62	35	8	93
13	Philco-Ford color composite, Apollo 9	Mar. 12	33	24	57	39	55	28	74	44	55	33	91
14	Multidate FRSL color composite (Pan–25A, high altitude)	March, May	76	86	3	----	----	----	----	74	29	0	100
15	Infrared Ektachrome, high altitude; images 7, 8, and 9 concurrently	March, April, May	81	82	4	78	37	75	17	83	21	35	72

Test image no.	Imagery type	Date	Total percent correct	Wheat		Bare soil, moist		Bare soil, dry		Bare soil		Cereal	
				Cor.	Com.	Cor.	Com.	Cor.	Com.	Cor.	Com.	Cor.	Com.
1	Pan–25A, Apollo 9	Mar. 12	43	0	100	63	62	54	31	83	29	39	33
2	IR–89B, Apollo 9	Mar. 12	47	0	100	42	35	82	41	89	17	42	27
3	Pan–25A, high altitude	Mar. 12	47	0	100	42	58	87	41	88	30	40	31
4	Pan–58, high altitude	Mar. 12	41	25	88	21	68	75	32	62	31	35	41
5	IR–89B, high altitude	Mar. 12	45	0	100	76	31	57	40	86	15	40	41
6	Infrared Ektachrome, Apollo 9	Mar. 12	65	50	83	83	25	75	20	98	3	83	27
7	Infrared Ektachrome, high altitude	Mar. 12	64	0	100	75	0	100	26	100	7	36	30
8	Infrared Ektachrome, high altitude	Apr. 23	60	0	100	----	----	----	----	96	10	66	22
9	Infrared Ektachrome, high altitude	May 21	72	0	100	----	----	----	----	81	20	87	9
10	FRSL color composite, high altitude	Mar. 12	58	0	100	75	37	84	23	100	11	37	19
11	FRSL color composite, Apollo 9	Mar. 12	50	0	100	58	41	63	48	97	14	43	19
12	Philco-Ford color composite, high altitude	Mar. 12	46	25	92	69	60	68	39	98	29	29	22
13	Philco-Ford color composite, Apollo 9	Mar. 12	33	50	67	23	74	56	42	42	54	27	54
14	Multidate FRSL color composite (Pan–25A, high altitude)	March, May	76	0	100	----	----	----	----	88	26	86	33
15	Infrared Ektachrome, high altitude; images 7, 8, and 9 concurrently	March, April, May	81	25	50	----	----	----	----	98	8	81	3

Cor = percent correct; Com = percent commission.

NOTE.—Space photographs, sequentially obtained aerial photographs, and corresponding color composites were examined by 12 photointerpreters. This table summarizes tables 3–7 through 3–15.

make the composite, the actual combination of bands used, and the enhancement filters chosen. So here, too, the particular composite used in the interpretation test described in this chapter may not be optimum; hence the reader is cautioned against accepting test results for composites as representing the optimum for interpretability of color composites.

Finally, the interpretation results from the 15 test images can be compared visually by graphing the percent of correct identifications for crop categories against the percent of commission errors (fig. 3-22).

In this graph, barley, alfalfa (both recently cut and mature), and bare soil (moist and dry soil combined) are represented as circles, squares, and triangles, respectively. Each number corresponds to the test-image number. A casual look at the graph shows that on most of the test images, barley was correctly identified between 25 and 60 percent of the time; three test images (15, 14, and 9) show barley correctly identified more than 80 percent of the time. Also, for most of the test images, interpreters incurred a 20- to 60-percent commission error in attempting to identify barley. For the alfalfa category, interpreters identified most of the fields between 45 and 80 percent of the time, incurring about the same commission error as for barley. Most of the bare-soil category was identified correctly more than 80 percent of the time; and commission errors were considerably less than for barley or alfalfa.

An example of what this graph indicates, then, is that test images obtained in March or April do not successfully discriminate barley fields (only 25 to 60 percent of the barley fields were identified). However, test images that consist of photos obtained in May (15, 14, and 9) were best for discriminating barley fields, yielding 80 percent or better correct identification for barley.

Improvement in identifying alfalfa (over that of barley) is attributable to the fact that recently cut alfalfa fields were for the most part readily identifiable. The commission error was high for alfalfa and barley partly because mature alfalfa and barley, in March, were frequently confused with each other. The most acceptable test image for identifying alfalfa was 15, the test in which March, April, and May Infrared Ektachrome

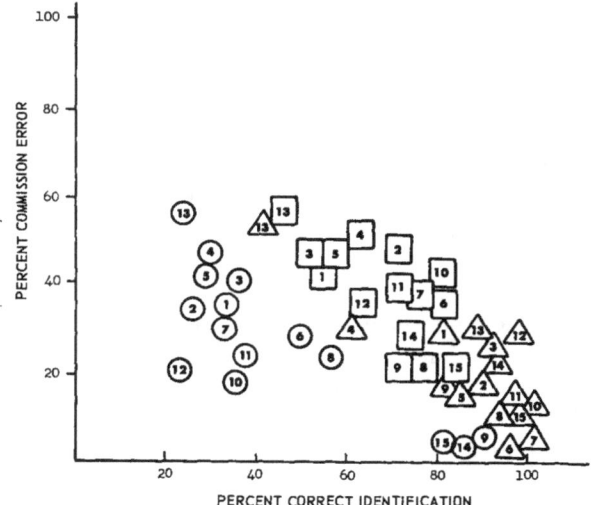

FIGURE 3–22.—This graph shows the relative interpretability of barley (circles); alfalfa, both recently cut and mature combined into one category (squares); and bare soil, both moist and dry combined (triangles) on each of the 15 test images examined in the interpretation test. Numbers pertain to the image types appearing in table 3–16. The reader should note in particular those test image numbers that permit interpreters to correctly identify crop categories with a high percentage of correct identifications, while incurring a low percentage of commission errors. Test image 15, for example, is the test image judged to be the most interpretable for all agricultural categories. Test image 15 represents the results for interpreting concurrently three high-altitude Infrared Ektachrome photographs obtained in March, April, and May, 1969. The data for preparing this graph came from table 3–16.

images were interpreted in concert. One might conclude from this that the changing pattern of alfalfa fields (from mature, to cut, to mature) is an aid to identifying it; hence, sequentially obtained photos (images) would understandably give higher correct identifications.

In the case of bare soil, most of the test images were acceptable for discriminating bare soil from all other categories of cover crops. It may also be of interest to note that those test images on which bare soil was identified with an accuracy of greater than 80 percent, with less than a 20-percent commission error, contain an infrared-sensitive film.

This graph also is useful for quickly showing those test images that provide an acceptable level of correct identification and an acceptable commission error for a crop category of interest to the ultimate user. If, as an example, the criteria for making an agricultural census were 80-percent-or-better correct identification and less than 20 percent commission error, then test image 15 probably would be of most interest.

Finally, and not too surprisingly, we find that agricultural categories are best identified on sequentially obtained images.

Subsequent to completing the test just discussed, a group of photointerpreters was asked to examine various combinations of Infrared Ektachrome prints from dates judged optimum for discriminating each major crop type. Different sets of photographs were used for identifying: (1) barley, (2) cotton, (3) grain sorghum (milo), and (4) alfalfa and sugar beets (one test for these two crop types). These dates were chosen by examining the signatures of crops on various dates and by using the crop calendar (fig. 3–18) to predict when changes in crop development might permit discrimination of particular crops. These photos included some not used in the previous test. Images taken on the dates indicated below were selected as being optimum for identifying the particular crop types for the reasons indicated:

(1) *April 23 and May 21.* Mature barley has a unique signature at this time of the year which permits the crop to be identified on photographs of these two dates (Color Plate 5, for example).

(2) *May 21 and September 30.* Two crop types, alfalfa and sugar beets, can be identified on these images. Two types of vegetation are distinguishable on a May 21 image: (1) bright-red color = mature alfalfa or sugar beets; (2) pinkish-gray color (various shades) = recently cut alfalfa. Confirmation of alfalfa versus sugar beets for the fields in the first category is made by checking these fields on a September 30 image to determine if the field is vegetated (bright red), in which case it is alfalfa, or nonvegetated, in which case it is bare soil in a field that was recently occupied by sugar beets but which now has been harvested and tilled. Color Plate 5 contains an example of alfalfa and sugar beets. The ground data maps for May 21 (fig. 3–4) and September 30 (fig. 3–8) can also be used to locate other such fields.

(3) *March 12 and September 30.* Most fields which contain bare soil in March will be planted to cotton later in the year. No other fields will be planted with this crop. Thus one can predict where cotton will be planted by locating bare soil fields (BS_d and BS_m) on a March 12 photograph. After the cotton crop is well established, one can check to determine if the predicted fields were actually planted to cotton. This is done by studying the predicted fields on a September 30 photograph and looking for the following: (1) bright-red color = vegetated field (cotton), (2) reddish-brown color = vegetated field (milo), and (3) light or dark gray (various shades) = bare soil. In this way, noncotton fields can be discriminated from cotton fields.

(4) *September 30.* Milo (grain sorghum) can be identified on an image from this date by its characteristic dark-red to reddish-brown color. The ground data from figure 3–8 can be used to locate examples of milo fields on the September 30 image in Color Plate 5.

Table 3–17 presents the results of the interpretation tests by each interpreter on the five crops discussed. Data from these tests were not as encouraging as expected. In every case, accuracy was slightly less than when some other image was interpreted, for certain crop types.

The results for one interpreter were far lower than for the other interpreters. If interpreters had been chosen on the basis of performance on tests of this type, perhaps results could have been improved. The same statement, of course, can be made with regard to other interpretations performed on images of a single date. Some interpreters commented, after studying their completed work with the correct answers at hand, that the training examples did not contain the full range of variability of the crop being identified. Even so, most interpreters seemed able to do a very satisfactory job. Acquiring greater familiarity with multiple-image tests of this type was also mentioned as a means of increasing accuracy of identification, because the interpreters had not worked extensively with multidate images in the past.

Detection of Crop Vigor

Some investigations are underway to determine the extent to which crop vigor can be determined using high-altitude photography. The most easily recognized evidence of vigor problems is that re-

TABLE 3–17.—*Interpretation Results From Analysis of Nikon Infrared Ektachrome Images Judged Optimum for Discrimination of Each Crop Type in the Mesa Test Site*

Crop type and interpreter No.	Number of fields	Image date	Correct responses		Commission errors	
			Number	Percent	Number	Percent
Barley	35	Apr. 23, May 21				
1			30	86	15	33
2			17	49	1	5
3			15	43	0	0
4			31	89	1	3
Average				66		15
Alfalfa	41	May 21, Sept. 30				
1			36	88	9	20
2			33	81	5	14
3			15	37	5	25
4			31	76	2	6
Average				70		16
Sugar beets	7	May 21, Sept. 30				
1			2	28	4	67
2			3	42	3	50
3			1	14	1	50
4			3	42	3	50
Average				32		55
Cotton	19	Mar. 12, Sept. 30				
1			11	58	1	8
2			17	90	1	5
3			17	90	3	15
4			19	100	2	9
Average				84		10
Milo	14	Sept. 30				
1			8	57	9	53
2			7	50	1	12
3			7	50	4	36
4			4	29	3	43
Average				46		39

sulting from the effects of soil texture, alkalinity, and fertility on crop growth. The effects of soil condition are usually quite evident because they produce characteristic patterns, are often extensive in area covered (can be resolved on small-scale photographs), and frequently cause distinct reduction in plant growth.

In Color Plate 9, nearly all the dark-red fields in (a) are citrus groves planted northeast of Mesa because of the region's favorable climatic regime. The meandering channel of the Salt River can be seen cutting across the upper-left portion of the photograph. Many orchards have been planted on the land adjoining the Salt River that was once part of the river channel. Old river meanders can be detected by recognizing the braided pattern that appears within certain orchards where tree growth is drastically reduced or where gaps occur. The arrow on the photograph in Color Plate 9(a) indicates the area where this condition is common. Low fertility of the gravelly soils of this area that affect tree growth can be contrasted with the more loamy soils in the lower part of this same photograph where orchard and crop growth is normal and this meander pattern does not occur.

The influence of soil type and landform on agricultural development can also be seen in Color Plate 9(*b*). The area imaged here is along the Gila River 20 mi west of Phoenix. The soils there consist of alluvial material washed down from the White Tank Mountains (off photograph at bottom). The stream channels that carry this alluvial material to the Gila River extend from the White Tank Mountains to the Gila River (top) and are evident on the photo. These channels can be seen both in the lower portion of the photo where they cut across wildland, and in the portion adjoining the Gila River where their pattern persists in agricultural land and is most apparent in bare cultivated fields. Three distinct soil-type boundaries can also be delineated on this photograph. They have been indicated in Color Plate 12 and the following soils of three associations are distinguishable: (1) old alluvial soils (Laveen-Whitlock Associations), highly calcareous, moderately coarse to medium textured (light-toned bare fields); (2) recent alluvial soils (Gila-Pima-Avondale Associations), moderately coarse to moderately fine textured (dark-toned bare fields); and (3) sandy to very gravelly river channel material (Arizo-Brazito-Vinton Associations). Further information regarding these soil types appears on the Soil Map of Central Maricopa County prepared by the Soil Conservation Service. The leaching of alkali accumulations was necessary in much of this area so that the land could be successfully farmed. A few fields at *A* on Color Plate 9(*b*) have not been reclaimed and show the effects of alkali accumulation.

High-altitude aerial photographs of the type studied in this report should be quite useful for determination on a regional basis of areas of low soil fertility as well as those areas that can be reclaimed and successfully tilled. Soil mapping on a regional basis could be undertaken in such a way that the relation of broad units of different soil type to each other might better be understood.

WILDLAND VEGETATION

The primary use of the wildland that surrounds Phoenix (see Color Plate 2) is for grazing by domestic and wild animals. The Bureau of Land Management and the Forest Service are the primary custodians of these lands. The native range vegetation consists of two basic types: the *semidesert shrub type* that occupies the alluvial plains to the north and southwest of Phoenix, and the *chaparral type* in the upland mountains to the north of Phoenix.

Within the semidesert shrub area the annual rainfall is low, the temperatures are high, and palatable forage production is poor. The shrubs that dominate these ranges are woody and provide little forage of value. Grazing is seasonal, occurring mainly in the early spring when annual grasses and weeds are present and in a green stage of development.

Chaparral ranges are located at higher elevation, and receive more rainfall. Considerably more of the chaparral shrub species are palatable, and the forage value of such ranges can be improved still further, either by removing some of the chaparral by fire or by mechanically clearing the area, and then seeding it to perennial grasses that will provide good grazing in the late-spring and summer season.

Our analysis of the wildland vegetation in this area as imaged on Infrared Ektachrome space photographs obtained by Apollo 9 confirms the usefulness of space photographs as a basis for mapping broad vegetation types. The two broad vegetation types of interest in the Phoenix area (semidesert shrub and chaparral) can be readily mapped because of their close association with recognizable landforms (alluvial plains and upland mountains, respectively), and their color signatures.

Frequently, more detailed information concerning the native range plants is desired (e.g., species composition, percent plant cover, density, distribution, and amount of forage material available). Because of the relatively low resolution of space photographs, they do not lend themselves to the direct extraction of this information. There is great promise, however, that information regarding important vegetation parameters can be obtained indirectly from the interpretation of space photographs by exploiting the "convergence of evidence," one example of which will now be given.

The diagram in figure 3–23 shows the close correspondence that exists in the semidesert shrub re-

50

VEGETATIONAL TYPES	YUCCA-AGAVE-SOTOL		PALO VERDE-SAGUARO-CACTUS		CREOSOTE BUSH	MESQUITE	WILLOW-COTTONWOOD	SALTBUSH
	YUCCAS OCOTILLO TURPENTINE BUSH PRICKLY PEARS FALSE MESQUITE	AGAVES SOTOL	PALO VERDES OCOTILLO PRICKLY PEARS EPHEDRAS JOJOBE	SAGUARO DESERT IRONWOOD CHOLLAS BRITTLE BUSH BUR SAGE	CREOSOTE BUSH BUR SAGE WHITE THORN CAT CLAW	MESQUITE CAT CLAW SALTBUSHES LYCIUMS JUJUBE	WILLOWS COTTONWOOD ARROW WEED BATAMOTE	SALTBUSHES GREASEWOOD PICKLEWEED

BASIN-AND-RANGE PROFILE

LAND FORMS & SOIL CONDITIONS	ERODING MOUNTAIN SLOPE	UPPER BAJADA	LOWER BAJADA	BOTTOM LAND	STREAM CHANNEL	SALT FLAT
	SHALLOW, ROCKY OR GRAVELLY SOIL WITH GOOD DRAINAGE NO SUBSURFACE WATER	COARSE-TEXTURED, ROCKY, WELL-DRAINED SOIL PARTLY UNDER-LAID BY ROCK BENCH. NO SUBSURFACE WATER	SANDY AND FINE-TEXTURED SOIL OFTEN WITH CALICHE HARDPAN. NO SUBSURFACE WATER.	FINE-TEXTURED SOIL WITH POOR DRAINAGE AND LOW SALT CON-TENT SUBSURFACE WATER AVAILABLE		SIMILAR TO BOTTOMLAND BUT WITH HIGH SALT CONTENT

gion around Phoenix, between landform, vegetation type, rock and soil type, and availability of subsurface water. Exploiting this relationship to the fullest, a photointerpreter can, for example, easily identify upland eroding mountains on a space photo or high-altitude aerial photo (Color Plate 10) and infer from this that soils, if present, are shallow and rocky; that the vegetation is sparse and for the most part unpalatable to livestock and game animals; and that subsurface water cannot usually be found. Using these same photographs, the interpreter can likewise readily detect and identify *bajadas* (coalesced alluvial fans). This knowledge then leads him to assume correctly that there is more vegetation (some of which is usable as forage) and more soil development, but only slightly more subsurface water than is found in the eroding upland mountains. Finally, in the bottomland areas (adjacent to large stream or river channels) the interpreter might expect to find both surface and subsurface water, a distinctive variety of hydrophytes (water-loving plants, many of which are palatable), and fine-textured soil that may be high in salt content.

In this instance, as in several others, the following important conclusion is indicated: even though there is not sufficient detail on space photographs to discern directly wildland vegetation parameters and soil conditions, these and many other attributes that relate to the potential productivity of the landscape can be inferred through the convergence-of-evidence principle, keying on those features that are discernible, namely, the landforms.

Native plants, like agricultural crops, progress through characteristic growth stages in response to favorable environmental conditions. Unlike agricultural crops, the wild vegetation must gear its development to the seasons when natural environmental factors (temperature, rainfall, etc.) are favorable. The time of growth and the length of the growing season of natural vegetation are not as predictable as those for crops. Nevertheless, these changes occur and valuable information can accrue through their detection, thereby facilitating the analysis of various wildland resources.

The chaparral vegetation that lies along the alluvial plains between Phoenix and the mountains to the north can be seen on the Infrared Ektachrome space photograph in Color Plate 11. On that photograph one can readily observe the variation in red, pink, and gray colors that characterize certain portions of the area. These rangelands are not considered highly productive in terms of the number of grazing animals that can be sustained, but, as in more productive areas, there is a high correlation here between the color response of the space photograph, area by area, and the amount of healthy vegetation on the ground at a particular season.

On the space photograph, the rangelands that had a conspicuous reddish color in March 1969 supported a dense cover of annual grasses and forbs (foxtail chess, *Bromus rubens*, is a dominant species). The annual plants germinate, develop, and mature in response to the amount and distribution of precipitation and temperature. Because the annual plants were nearly at a peak stage of growth when the space photo was taken, one can readily identify the areas in which they are found. In contrast to the reddish areas that had the greatest volume of healthy plants, the gray areas (without red or pink) had the least healthy vegetation at the time the photograph was taken. This last point is worth further amplification because in any given region, physiographic variations (elevation, slopes, and aspects) create small environmental changes that cause plants of the same species to develop at slightly different times. Hence,

FIGURE 3–23.—The diagram below the photographs, from Benson and Darrow (1954), indicates the close association between vegetation types, landforms, and soil conditions that exists throughout much of the arid southwestern United States. It has been documented that many of the major plant species associated with the various landforms, as shown in the diagram, also occur on the corresponding landforms that appear in the area seen in Color Plates 10 and 12. The photographs above were taken on the ground at the indicated profile positions. The species that appear in these photographs are the same as those described in the text. The significance of this relationship of vegetation and landforms is apparent when interpreting space photographs, such as the one in Color Plate 10, where the vegetative detail is lacking, but the landforms are conspicuous.

when our interpretations suggest that an area has a lesser amount of vegetation, it should be remembered that this condition may exist only at that particular time, either because the vegetation has already developed and died, or because it is late in developing and is not yet manifest on the photo. Thus we need sequentially obtained images to improve the validity of our analysis of time-variant phenomena.

The aerial photo (part of a high-altitude panoramic photograph) in the lower left of Color Plate 11 shows the same area as that in the rectangle outlined on the space photo and was taken on the same date. The correspondence between the reddish color and the amount of healthy vegetation occupying specific range sites can be readily seen. Note the area that has considerably less vegetation in the upper-right corner of the aerial photograph (see *A* on March 12 aerial photograph, Color Plate 11). This may be related to differences in soil type or associated with surface erosion. Notice also that cultural features such as canals, roads, small erosion scars, and drainage channels are seen in greater details, but that interpretations of the amount and distribution of vegetation are made as well on the space photograph.

The lower-right aerial photo in Color Plate 11 shows essentially the same area as the one on the left, but was taken on April 23, 1969. The absence of the conspicuous red areas where they had existed earlier in March indicates that the annual plants had completed their short life cycle and died. Sequential viewing of such an area can provide information as to when the annual crop begins to develop, when it is at an optimum growth stage for grazing (a condition known as "range readiness"), how long the crop remains green, and the distribution of the crop within a large range area. In this manner, grazing activities can be more efficiently synchronized within a region to utilize the available forage resources more fully. Note on the aerial photograph taken in April that the density of perennial plants is not high enough to indicate the amount of this vegetation. Also note that soil-type differences are not as readily detected.

Color Plate 12 shows another sequence of aerial photographs. These were taken along the Gila River just south of Phoenix, and cover the same area seen in Color Plate 10. They illustrate how changes in the phenology of certain plants can give clues to the physical characteristics of an environment. For example, one can see three conspicuous terrain types within the river-channel environment. They are characterized in the March 12 photo as follows: Light-gray areas (*A*) contain the highest concentrations of salt and support the least vegetative material, consisting of only the most salt-tolerant species; pink-red areas (*B*) indicate a cover of annual plants (grasses and forbs) on a site also high in salt content. The difference between the light-gray and the pink-red areas is attributed to the dense cover of annual plants in (*B*). That annual plants are better able to grow in area (*B*) indicates a lower salt content and perhaps differences in soil texture. Notice that areas (*A*) and (*B*) are best differentiated on the March photograph, but the distinction between these two areas is lost or easily overlooked at subsequent dates. Conspicuous dark areas (*C*) are stands of mesquite and tamarisk. Their presence indicates a subsurface ground water supply. Note that on the April photograph both species appear red because both have leafed out. The individual species cannot be differentiated at this date. On the August photograph, however, the tamarisk is still in leaf and appears red, whereas the mesquite has begun to lose its leaves and appears dark gray again. Tamarisk forms dense stands immediately adjacent to the stream channels that flow intermittently during the year. Mesquite tends to grow in more open stands, farther back from the immediate vicinity of the stream channels. The significance of being able to identify water-loving species and to determine when they are in or out of leaf is related to the tremendous quantity of water loss (due to transpiration) by such species. Thus, in such areas, the presence of abundant subsurface water can be inferred; and judicious removal of such species, without disturbing the stability of the stream channel, can reduce the loss of water by transpiration.

Plants that indicate the level of salt in the soil are also very important when reclamation of land for agricultural development is considered. Note in Color Plate 12 that a few farmers have tried to cultivate land adjacent to the river bottomland. Some farmers apparently were successful, whereas others abandoned the effort because of the high salt concentrations.

THE GEOLOGIC AND HYDROLOGIC RESOURCES

In connection with the S065 experiment, FRSL was asked to evaluate all Earth resources (including geologic and hydrologic resources) in the area centered around Phoenix, Ariz., which was covered by the S065 photography. This section summarizes the geologic and hydrologic evaluations completed to date in response to that request. The primary responsibility for evaluating geologic and hydrologic resources rests, however, with the U.S. Department of Interior Geological Survey; persons interested in evaluations of geological and hydrological resources should consult the work done by the U.S. Geological Survey and Dr. Paul Lowman of NASA, principal investigator for the S065 experiment.

Geology

Apollo 9 Photography

Because the scale and resolution of the Apollo 9 photography are significantly different from conventional photography, the Gila Bend region of Arizona was selected as an area with which to familiarize the interpreter with the special aspects of geologic and geomorphic features as seen on Earth orbital photography. Only after such familiarization could efforts be made to determine the interpretability of geologic and geomorphic features on Apollo 9 photos of other areas.

During this familiarization study it became apparent that the photogeologist must pay attention to subtle differences in image characteristics of various geologic features when interpreting and evaluating orbital photography of such small scale. By means of a geologic overlay of the Gila Bend region, using the Arizona Bureau of Mines Geologic Map of Maricopa County as reference, geologic features found in this area were studied as to their image characteristics on Infrared Ektachrome photography (Color Plate 13).

The Gila Bend area contains a variety of rock types, including volcanic, intrusive igneous, metamorphic, and sedimentary rocks. Much of the area, however, is covered with unconsolidated quaternary alluvium, and the various rock types occur as bedrock islands within this alluvium. Lithologic distinctions within the bedrock areas can sometimes be made on the space photos using tonal and textural differences. For example, andesite in this area appears reddish-brown in color on the Infrared Ektachrome print and has a highly dissected topography which appears as low angular hills. As seen on Color Plate 13, these image characteristics contrast with the darker tones of the basalt where it is not masked by vegetation. It will also be noticed that the basalt is less eroded than the andesite and forms longer, higher ridges or isolated hills, or occurs as extensive flat areas. (Low-altitude oblique photos showing ground conditions in this area are seen on Color Plate 14.) The granitic rock appears light in tone and differs in topographic form (generally forming long, thin, acutely branching ridges) from the volcanic rocks present. Granitic gneiss, however, cannot be easily differentiated from the granite because both exhibit a similar photographic tone and topographic form at this scale. Despite the tonal and topographic differences exhibited by the various rock types on these space photographs, boundaries between lithologic units are not always distinct.

Using experience gained from the study of the Gila Bend area, follow-on studies were made in the Roosevelt-San Carlos area east of Phoenix (fig. 3-24). These studies were carried out to test the degree of success obtainable in geologic mapping from orbital photography at the scale (1/3 000-000) and resolution (250 ft) presently available. For purposes of this test, the Apollo 9 photography was enlarged to a scale of 1/250 000. The relative usefulness of the various film-filter combinations used in the S065 experiment was also tested.

The area between Roosevelt and San Carlos Reservoirs was chosen for geologic mapping because it appeared to have a variety of geologic features that could be mapped from the space photos. Also, the geology of the region was totally unknown to the interpreter at the beginning of the study, and thus it was a good area for objective study. The area had been covered by high-quality S065 photographs, which included Infrared Ektachrome and three black-and-white bands (Pan-25A, Pan-58, and IR-89B). However, because only 5 to 10 percent forward lap was obtained, stereo coverage was inadequate. Conventional Ektachrome photography of this area was not obtained on the Apollo 9 mission.

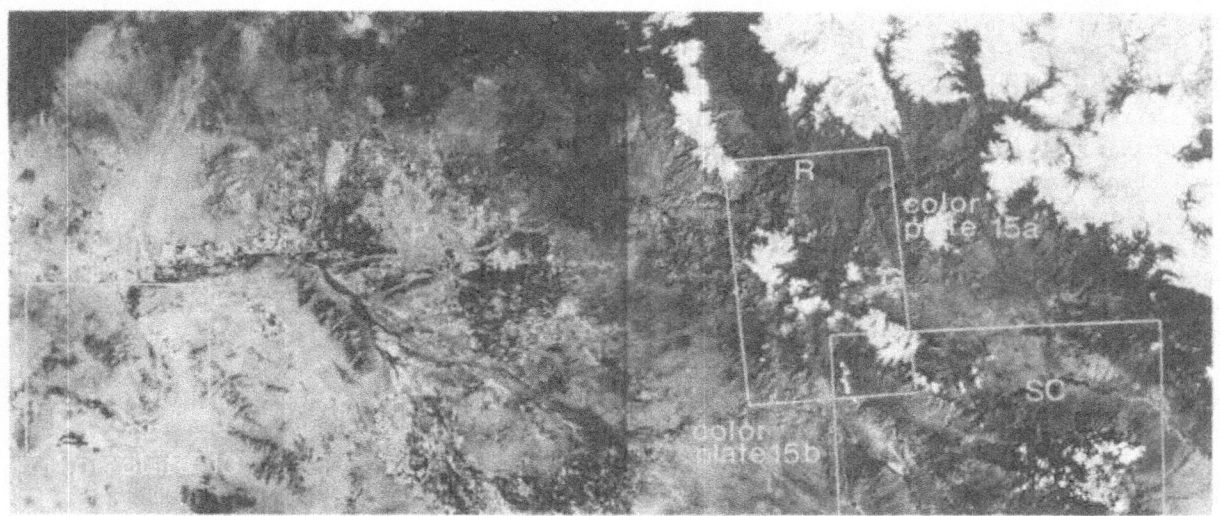

Legend

G   Globe
GB Gila Bend
M   Mesa
P   Phoenix
R   Roosevelt Reservoir
SC San Carlos Reservoir

FIGURE 3–24.—Photomosaic compiled from black-and-white prints of Infrared Ektachrome Apollo 9 photographs of the Phoenix area showing the areas for which the geologic evaluation was carried out. The boxes indicate the areas shown in Color Plates 13, 15(*a*), and 15(*b*).

Ground information in the form of existing geologic maps is available for the area. However, with the aid of experience gained from the study of the Gila Bend area, the Roosevelt-San Carlos area was initially interpreted without reference to ground data.

Portions of the original 70-mm transparency were enlarged and made into 8- by 10-in. prints (approximately 10 times the original scale) and into 2¼- by 3¼-in. transparencies (approximately three times the original scale) for the purposes of the study. The features delineated consisted of homogeneous tonal, textural, and topographic units that were discernible on the photos. A separate photo map was made for each film-filter combination. After delineation of the geologic units in the designated area was completed on all the film-filter combinations, the maps were compared with the existing geologic maps of the area.

An example of the results of this study can be seen in figure 3–25 and Color Plate 15(*a*). These figures show a part of the study area (with photo-delineations for the different film-filter combinations) as well as the ground data for the area. As can be seen, there is a good deal of discrepancy between the maps derived from the interpretation of the space photos and the ground data. Most easily delineated were areas of recent alluvium. This was true for all of the film-filter combinations. Tailings and tailing pond areas around active mining and processing centers such as Miami, Superior, and Sonora could also be easily delineated on all but the black-and-white infrared photography taken with an 89B filter. The tailing ponds and tailings which register as light gray to white on the black-and-white photography were most accurately delineated on the Infrared Ektachrome photo where they registered as a light-blue tone that made it easy to distinguish them from clouds and small snow patches.

In areas of bedrock, the lithologic type most readily identified was quaternary basalt in the form of flows. The large area of quaternary basalt northeast of San Carlos Reservoir (fig. 3–24) was accurately identified on all of the film-filter combinations. Smaller areas of basalt were not recognized as such on any but the Infrared Ektachrome photographs. The large volcanic area southwest of Superior was identified correctly on the Pan-25A and Infrared Ektachrome space photographs, whereas on the Pan-58 and the IR–89B photographs there was ambiguity as to whether the area was occupied by volcanic or metamorphic rock types. A large intrusive igneous (granite) area north of Apache Reservoir was correctly identified in the Four Peaks region; however, the full extent of its boundaries was not correctly delineated on any of the photos. A large area south of Miami which was interpreted as intrusive igneous rock is part granite and part schist. The area identified as instrusive igneous rock north of Superior is actually schist. Thus, there was a question about what was intrusive igneous rock and what was schist (metamorphic rock) in this area. Part of this problem may be eliminated by further training of the interpreter. A large area east of Sonora was correctly identified as sedimentary rock from the expression of bedding in the topography; however, it was impossible to identify smaller areas of sedimentary rock where bedding was not sufficiently resolved.

Many boundaries could be delineated in part by tonal variations such as the light, elongate area south of Roosevelt Reservoir, but such areas could not be identified as to lithologic type at this scale.

There are some interesting tonal variations within the quaternary deposits just north of San Carlos Reservoir (fig. 3–24) which in places correlate with lake deposits in that area. Other tonal variations within the alluvium of that area cannot be immediately evaluated without extensive field work. Many of the tonal variations are definitely related to variations in vegetation associations. Perhaps some of these vegetation associations can be correlated with specific geologic conditions. Thus there is need for further research to test the value of these space photos for the study of quaternary geologic processes and environments.

In summary, the Infrared Ektachrome photos were the most interpretable for discriminating geologic features by virtue of the distinguishable color differences. Of the three types of black-and-white photography, Pan-25A proved to be the most useful. Since most geologic interpretations were made on the basis of topographic expression of any particular geologic condition, the wider range of tonal contrast of the Pan-25A photos as compared with the other black-and-white photos made them the most easily interpreted. Geologic features had the lowest range of tonal contrast on the Pan-58 photos. These were the hardest to work with, partly because they were underexposed.

Infrared–89B photos were intermediate in tonal contrast between the Pan-25A and Pan-58 photos. However, the IR–89B photographs appeared to have higher resolution than the other film-filter combinations. This may in part have been due to techniques used in the reproduction of the second-generation copies of the original transparencies.

The study of geologic features in the geologically complex area east of Phoenix suggests that for a geologist with only moderate interpretation experience, satellite photography having a scale of 1:3 000 000 and a resolution of approximately 250 ft is adequate for only gross delineations of large lithologic units and relatively large simple structures. Studies of Apollo 6 photos of Texas by more experienced photogeologists (Amsbury, 1969) support the same findings:

> Many formation boundaries shown on the geologic map of Texas (1933) with a scale of 1:2,000,000 cannot be seen on the Apollo 6 photographs; and many faults shown on the Dallas quadrangle map (1968) with a scale of 1:250,000, cannot be seen. In more complex areas such as near El Paso, little information about the bedrock geology can be obtained from the photographs. A resolution of 100 feet to 100 meters is evidently not sufficient for geological study at a scale of 1:2,000,000 much less at a scale of 1:250,000. . . .

However, in areas of less-complicated geology than the test area, a higher degree of success could be expected, based on findings in the Gila Bend area. In areas of recent sedimentation, such as the vast areas of valley fill and in the larger stream channels, definite tonal variations could be seen on the photographs that might relate to variations in the sedimentation process and environment. Thus, space photographs of this scale and resolution might be more useful for studying contemporary processes such as in the quaternary deposits.

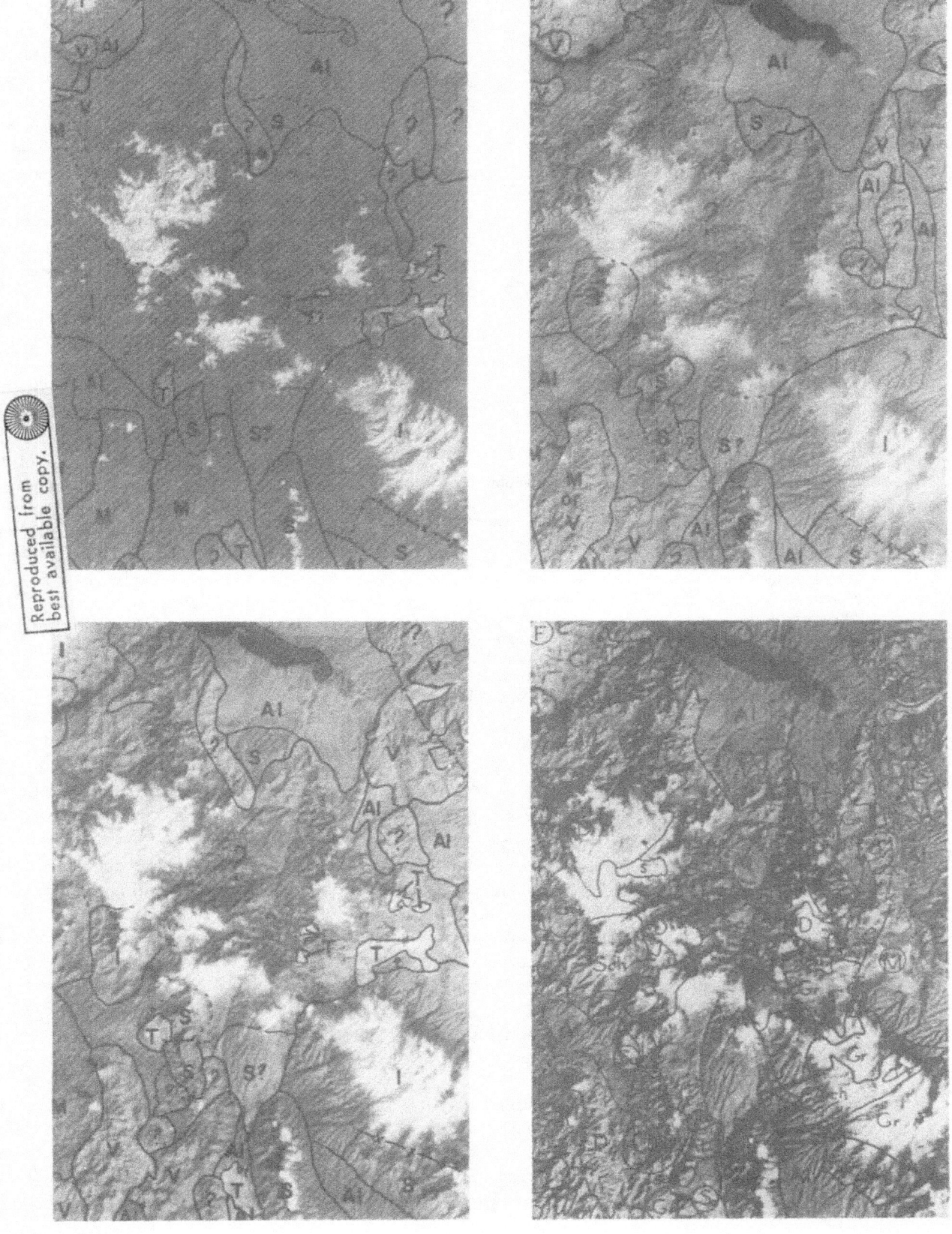

Legend

Areas (circled letters):
- A   Apache Reservoir
- F   Four Peaks
- M   Miami
- R   Roosevelt Reservoir
- So   Sonora
- Su   Superior
- T   Tailings and tailing ponds

Faults: –X–X–X–

Rock Types:
- Volcanic: V
  - An   Andesite
  - B   Basalt
  - D   Dacite
- Intrusive Igneous: I
  - Gr   Granite
- Metamorphic: M
  - Sch   Schist
- Sedimentary: S
  - Al   Alluvium

FIGURE 3–25.—An example of geologic mapping on Apollo 9 photos in part of the Roosevelt-San Carlos study area. Photographs *A*, Pan-58; *B*, IR–89B; and *C*, Pan-25A show the interpretations for the three bands of S065. Photo *D* is a black-and-white print of the Infrared Ektachrome with ground data delineations. Dotted lines indicate ambiguity in the delineations. Question marks within delineated units indicate that although the unit was delineated, the interpreter was unable to identify its lithologic type. See text for a fuller discussion of the results. (For interpretation of the Infrared Ektachrome photograph, see Color Plate 15(*a*).)

◀————

High-Altitude Photography

As a sequel to this study of space photography, an evaluation of high-altitude aerial photography was undertaken to ascertain the amount of added information that could be extracted from this type of imagery with its larger scale and higher resolution. The area of study was covered by both the high-altitude aerial photography and a portion of the Apollo 9 imagery, so that a comparison of the two types of imagery could be made.

The study area is a northeast-southwest-oriented strip over Aravaipa Canyon and the eastern end of San Carlos Reservoir (Color Plate 15(*b*)). Imagery of this area taken from an altitude of about 70 000 ft with 35-mm Nikon cameras having focal lengths of 21 mm was available in three black-and-white bands (Pan-25A, Pan-58, and IR–89B) and in Infrared Ektachrome. With approxi-

mately 70 percent forward lap between the photographs, there is good stereo coverage of the study area by the high-altitude photography. Based on findings from the Apollo 9 study, which showed Infrared Ektachrome to be the most valuable of the four types of imagery available for the extraction of geologic information, only this imagery was used in evaluating the high-altitude photography.

The original intent was to determine whether geologic interpretations could be facilitated by using high-altitude photos taken on two or more dates. While most geologic environments themselves do not change appreciably in a short time, it may be useful to monitor, at least through one season, changes in other factors such as vegetation that contribute to the interpretation of geologic features. At the least, the one best date to obtain high-altitude photography for geologic evaluation of the study area might have been determined. However, of the five dates of photography available to the interpreter for this area (July 15, August 5, August 29, September 30, and November 4, 1969), only the September 30 photography could be used in the study. The other dates were not suitable because of heavy cloud cover or (in most cases) improper exposure. For this reason no geologic evaluations of the sequential aspects of the high-altitude aerial photography could be undertaken. HyAc (panoramic) photography of the area taken simultaneously with the 35-mm photography, but with a camera having a focal length of 12 in., was received for only one of these dates (August 5). The area was obscured by heavy clouds on that date, so no HyAc photography was evaluated.

For purposes of the evaluation, duplicates of the original 35-mm photography with an approximate scale of 1:920 000 were enlarged to a variety of scales to determine the most beneficial scale with which to work. It was found, for work with a stereoscope possessing 8 × and 2.5 × lenses, that an 8 × enlargement of the 35-mm photography was most suitable. Such an enlargement results in an approximate scale of 1:114 000. The 8 × enlargement viewed under the 8 × lens of the stereoscope gives a workable scale of 1:14 250. During the interpretation, smaller-scale photographs were often referred to in order to interpret large geologic features more easily. One of the advantages of having small-scale, high-resolution photography of an area is that a variety of scales can be derived from a single flight. Within the limits of the available resolution, this leads to more complete interpretation. Estimates of the vertical exaggeration on these photos indicate that it is approximately 2 ✕.

Ground information was obtained from the 1 : 500 000 scale Geologic Map of Arizona. The interpreter also had visited accessible areas in the Aravaipa Canyon region and areas south of San Carlos Reservoir.

Color Plate 16 and figure 3–26 show portions of the study area which were selected to illustrate significant features discernible from the high-altitude photography. Figure 3–26 displays a sequence of Tertiary silicic volcanic flows. Several individual flows within the sequence can be distinguished on the high-altitude photos. The volcanic beds form a hummocky plateau that gently dips to the northwest. Most of the volcanics in this area appear as a light-pink tone on the Infrared Ektachrome photos, but a few are darker and have a violet tone. These dark beds are probably less silicic than the lighter toned beds. The volcanics exhibit a medium-to-coarse-textured, angular drainage pattern, and are overlain in places by Tertiary-Quaternary sediments consisting of loosely consolidated gravel, sand, and silt. Sharp, definite boundaries cannot be drawn between the volcanics and the Tertiary-Quaternary sediments because the two units have similar tonal characteristics. However, a boundary can be inferred from the analysis of the drainage patterns. The drainage pattern in the sediments is a fine-textured dendritic-pectinate (featherlike) pattern, whereas the drainage pattern in the volcanics is a medium-to-coarse-textured angular pattern. The boundary between the two units is tentatively placed between the two different drainage patterns. Limited field observations in the area indicate that the volcanics underlie the sediments along Aravaipa Canyon. The narrowness of the canyon and the steepness of its walls indicate this as a possibility.

The area shown in Color Plate 16 contains an isolated asymmetric syncline in sedimentary and volcanic rocks. Disparities in degree and direction of dip of discernible beds south of the synclinal structure indicate the presence of two major faults or unconformities. North of the syncline, several light-toned beds within the Tertiary volcanics are also visible and can be used as marker beds in the delineation of structure. Offsets and disruptions within these beds show the presence of another fault. Several linear features cutting across the axis of the fold and possibly related to fracturing or faulting can also be delineated on the high-altitude photography.

Compare the high-altitude photographs in Color Plate 16 and figure 3–26 with the Apollo 9 satellite photograph in Color Plate 15(b). The individual gently dipping flows visible on the high-altitude photograph in figure 3–26 cannot be separated on the Apollo photograph, but even more important, the lithologic identification is much more in doubt on the lower resolution satellite photos. However, when moderately tilted volcanics occur in conjunction with tilted consolidated sedimentary rocks, the two rock types cannot be separated on the high-altitude photo either. Also, a fault west of Aravaipa shows equally well on both the Apollo 9 and the high-altitude imagery. On the Apollo 9 imagery in the region of the syncline, a fold structure is vaguely indicated; however, it is not definite. Part of the problem on the Apollo 9 photograph is that scattered clouds had obscured the picture in a crucial area. However, if the Apollo photo were unobscured, the nature of the fold, i.e., whether it was an anticline or a syncline, still could not be determined. The high-altitude photo, on the other hand, clearly shows the presence of a fold; and since individual beds can be resolved and their direction of dip determined, it can be established that the fold is a syncline. The linear features that cut across the axis of the syncline and are possibly related to fracturing or faulting are not clearly visible on the Apollo 9 photograph. Of course, the Apollo photographs do not provide stereo coverage for this area, and one would expect to gain more information from them if stereo coverage were available. However, it is felt that high-altitude photographs offer the interpreter more information than can reasonably be expected from satellite pictures of this resolution even with stereo coverage.

In conclusion, high-altitude photographs can provide an adequate means of performing reconnaissance-type geologic mapping projects. The degree of success will vary, of course, depending on the complexity of the geologic situation, the density of vegetation cover, and other variable local condi-

59

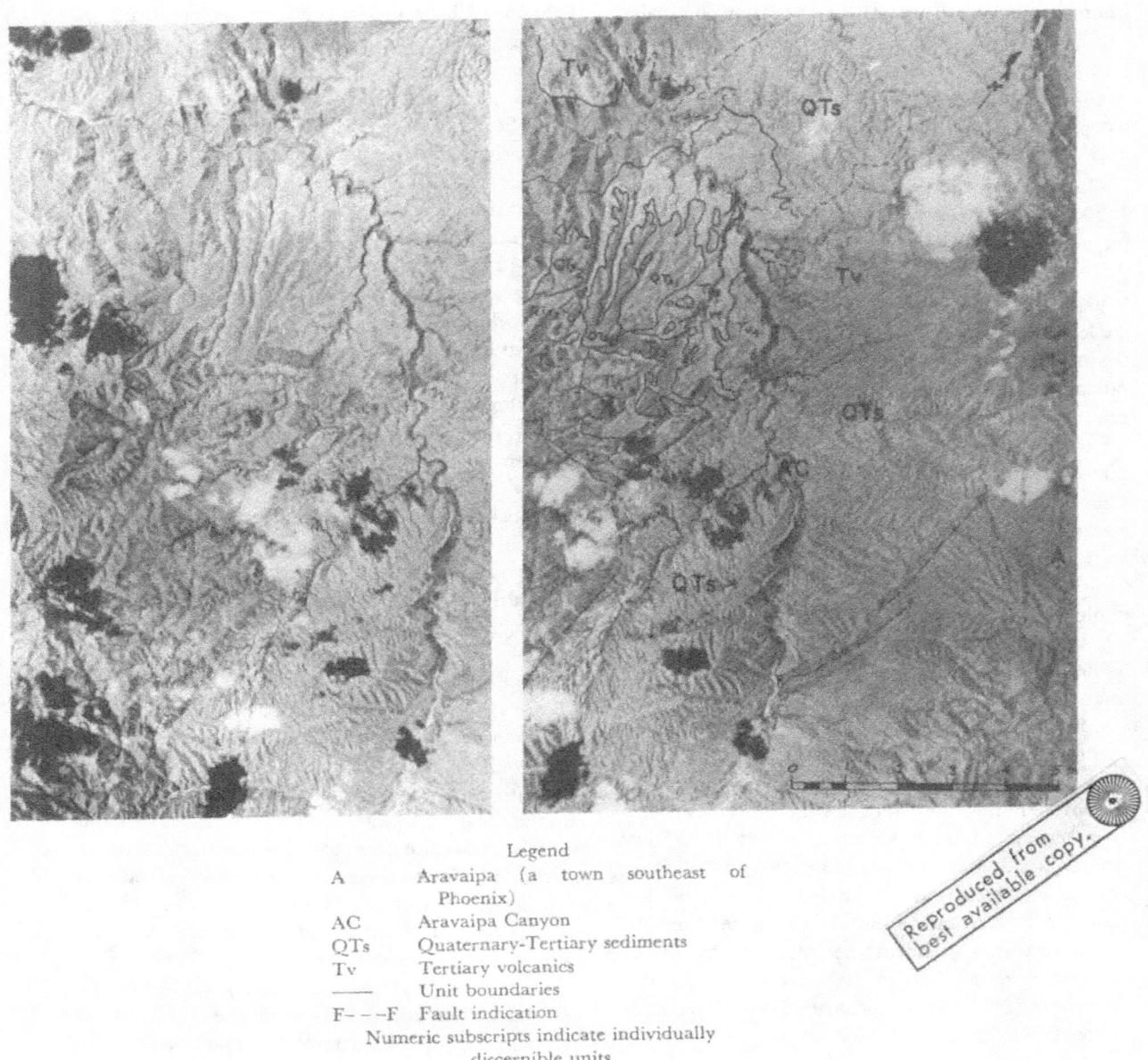

Legend

A	Aravaipa (a town southeast of Phoenix)
AC	Aravaipa Canyon
QTs	Quaternary-Tertiary sediments
Tv	Tertiary volcanics
——	Unit boundaries
F---F	Fault indication

Numeric subscripts indicate individually
discernible units

FIGURE 3-26.—This high-altitude stereo pair contains a sequence of volcanics that form a
hummocky plateau (Tv). The high resolution of the photography makes it possible to
delineate several of the individual flows within the sequence. This cannot be done on the
Apollo 9 photograph seen in Color Plate 15(b). Bordering the volcanics are loosely consolidated Quaternary-Tertiary sediments. These sediments have tonal characteristics
similar to the volcanics and so can only be differentiated from the volcanics by drainage
pattern analysis. See text for a fuller explanation.

tions. The combination of small-scale and relatively
high resolution offers greater flexibility than is presently available with satellite photography. The relatively higher resolution of high-altitude photographs offers added details needed for making
geologic interpretations with acceptable confidence
levels. Higher resolution also means that a wider
variety of scales is available to the interpreter, thus

offering the photogeologist the benefits of small-scale (synoptic view) photography, as well as medium scales for more detailed work. Certainly for very detailed geologic mapping, the use of large-scale, low-altitude photography, coupled with adequate fieldwork, still constitutes the preferred method. However, for the rapid reconnaissance of geologic resources, high-altitude photography is much better than the space photographs obtained by the Apollo 9 astronauts.

Hydrology

In many areas of the West the bulk of the annual runoff occurs in the spring and summer months with the melting of the winter snowpack from the high mountainous reaches of a drainage basin. It is desirable and often quite necessary to estimate the amount of seasonal runoff that is to be expected from a given drainage basin to plan for flood-control measures, irrigation needs, and other water-use requirements. Users with low-priority water rights find this kind of information vital to the planning of their activities. In areas where time and funds are not available for an extensive snow survey, it may be feasible to use space photography coupled with a limited on-the-ground sampling program to survey the season's snowpack and arrive at a usable estimate of the expected seasonal runoff. With the present state of the art, space photography would be used primarily to delineate snow-covered areas (fig. 3–27) and, through a comparative analysis of sequential space photographs, to monitor the rate of snow accumulation and snow melt. Depth and water content of the snow would be measured on the ground, but only at a few spots selected from a study of the space photographs.

To test the feasibility of using space photography to trace out drainage basins and thus assign each part of the snowpack to its receiving reservoir, the snowpack of the Four Peaks area in the Mazatzal Mountains, northeast of Phoenix, was selected for study. From interpretation of the space photograph, and without the aid of existing topographic maps, the snowpack was sectioned off in terms of the reservoir into which its melt waters would drain. Such information lends itself to efficient planning of water storage and release from each reservoir in a water-control system. The same delineations were made on the 1:250 000-scale topo-

graphic map of the area to check the accuracy of the delineations made totally on the photos. The results of this work can be seen in figure 3–27. The hatched areas indicate the portions of the drainage basin over which no melt water will drain. As can be seen by comparing the Apollo photograph with the topographic map, there is a discrepany between them as to the area over which no melt water will drain to the Saguaro Lake Reservoir and the Apache Lake Reservoir. The proportioning of the snowpack on the two photographs agrees fairly well, however. More reliable delineations could possibly be made with increased resolution plus adequate stereo coverage; however, this aspect of the study will have to wait until suitable imagery is available.

LAND-USE PATTERNS AND CULTURAL DEVELOPMENTS

Although the wildland and agricultural resources were given prime consideration in this study of the Phoenix area, the potential of space photographs for evaluating urban development, land-use patterns and other cultural developments was also recognized. Specifically, the large urban-industrial area of Phoenix, and many of the smaller surrounding cities (Tempe, Mesa, Chandler, and Casa Grande), are readily detected on Infrared Ektachrome space photographs. (See Color Plates 2 and 3.) The unique color signature of urban areas permits easy detection and determination of size. Major asphalt thoroughfares also have a unique appearance and can be seen in the towns of Phoenix, Tempe, and Mesa. Within the boundary of Phoenix, the number and distribution of parks, golf courses, and other open or natural areas that are vegetated can most readily be seen on the Infrared Ektachrome space photos.

The IR–89B space photographs also show the asphalt thoroughfares of downtown Phoenix, Tempe, and Mesa. However, the urban areas and their boundaries cannot easily be distinguished on the IR–89B photographs. Dirt roads, airports, land clearing for subdivision development, large farms, and feedlots can be detected more readily on Infrared Ektachrome, Pan-25A, and Pan-58 (in that order), but these features are not easily detected on the IR–89B space photos.

The benefits from obtaining a synoptic view of

61

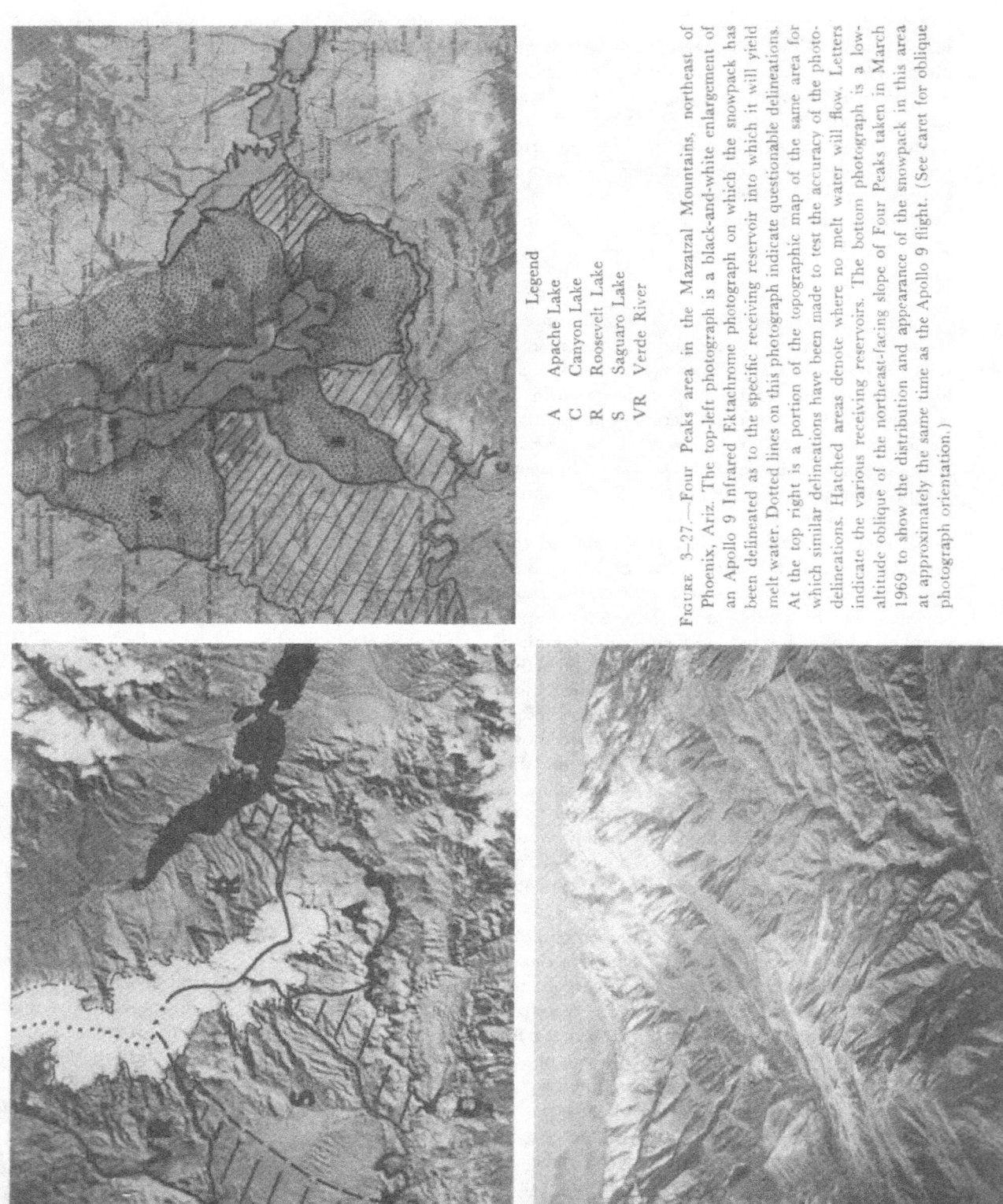

Legend

A   Apache Lake
C   Canyon Lake
R   Roosevelt Lake
S   Saguaro Lake
VR Verde River

FIGURE 3–27.—Four Peaks area in the Mazatzal Mountains, northeast of Phoenix, Ariz. The top-left photograph is a black-and-white enlargement of an Apollo 9 Infrared Ektachrome photograph on which the snowpack has been delineated as to the specific receiving reservoir into which it will yield melt water. Dotted lines on this photograph indicate questionable delineations. At the top right is a portion of the topographic map of the same area for which similar delineations have been made to test the accuracy of the photo-delineations. Hatched areas denote where no melt water will flow. Letters indicate the various receiving reservoirs. The bottom photograph is a low-altitude oblique of the northeast-facing slope of Four Peaks taken in March 1969 to show the distribution and appearance of the snowpack in this area at approximately the same time as the Apollo 9 flight. (See caret for oblique photograph orientation.)

urban, agricultural, and wildland areas accrue from the information that can be extracted for land-use planning. Attention is directed to Color Plate 2 wherein the broad land-use categories have been mapped from interpretation of the space photograph. Note, for example, that land already under cultivation is readily identifiable. Hence, the area and distribution of major crop types can be mapped, and the distances to a market center can be computed. Likewise, wildland that could become potentially productive agricultural land can be mapped and the best transportation routes and irrigation networks can be planned in advance.

Recreation planners can also use space photographs to locate potential recreation sites in wildland areas. Availability of water and distance from population centers may be important factors influencing the decision to develop recreation areas. This information can be obtained from interpretation of space photographs.

Information regarding water-supply needs, both for domestic and agricultural use, can be obtained from space photographs. Factors that can be evaluated pertaining to water-supply needs include the following: size and distribution of urban areas and agricultural land, distance from user to reservoir and watershed, number of reservoirs, reservoir levels, snowpack and distribution, and watershed characteristics that affect water yield.

During the course of our studies at the agricultural test site located just south of Mesa, Ariz., we have observed wih interest the extension of suburban development into prime agriculture land. Even within the agricultural test site new housing construction is underway, and the trend seems to indicate that more and more of the agricultural land will soon be converted for housing or mobile-home sites. Sequentially obtained photographs promise a means for monitoring the conversion of agricultural land to suburban development. Information derived in this manner will benefit municipal planners and utility companies in making advance plans for the extension of their services to meet the needs of the expanding suburban areas.

Prof. Duane Marble of Northwestern University is now making detailed studies of the extent to which urban-area analysis and land-use studies can be made on the Apollo 9 and associated high-altitude photographs of Phoenix and environs.

SUMMARY AND CONCLUSIONS

Earth resources in and around Phoenix, Ariz., that have been studied on Apollo 9 photography and sequential high-altitude photography include the agricultural, range, geologic, hydrologic, and cultural resources. The purpose of this study has been to determine the usefulness and/or limitations of such photography for evaluating and monitoring Earth resources.

Our analysis of agricultural resources of the Phoenix area produced these results:

(1) The great value of the crop calendar concept for describing patterns of crop sequence was shown. The crop calendar that we devised for the Phoenix area was used successfully to predict the optimum dates for identification of the major crop types in that area.

(2) Negative transmission values (i.e., values for the transmission of light through photographic negatives) were measured for representative fields in the Mesa test site. Such values were used successfully in certain instances to predict cases in which specific crop types might easily be discriminated on the basis of their tone values on panchromatic photographs.

(3) Multiband and multidate color composite images prepared by optical and electronic systems were compared. In some cases the identification of a particular crop was facilitated through the interpretation of a specific enhanced image.

(4) Agricultural studies determined that acceptable estimates of crop area can be made from the space photographs (in particular the Infrared Ektachrome, Ektachrome, and Pan-25A photographs).

(5) Crop identification to an accuracy exceeding 60 percent could rarely be made on the space photographs obtained on March 12, 1969, mainly because barley, mature alfalfa, wheat, and sugar beets have similar spectral signatures in March, making accurate discrimination difficult.

(6) The similarity of test results for crop identification using Apollo 9 space photographs and high-altitude photographs taken simultaneously on March 12, 1969, suggests that conclusions reached from analysis of sequential high-altitude photographs should apply equally well to studies of space photography obtained on a sequential basis. The improved resolution of the high-altitude photo-

graphs was not sufficient to improve interpretation results significantly.

(7) For each date of photography tested, higher interpretation accuracies were obtained from Infrared Ektachrome film than from Pan-25A film.

(8) The overall accuracies obtained from concurrent interpretation of March 12, April 23, and May 21 Infrared Ektachrome photographs were higher than from the separate interpretation of photographs from each of these dates.

(9) Certain soil-type boundaries and the effect on crop growth of certain soil conditions were easily detected in high-altitude photographs. The potential for expanding studies of this type was indicated.

Wildland vegetation resources were also evaluated on the space photographs, placing particular emphasis upon the forage resource. Infrared Ektachrome photographs proved to be the most useful for this purpose, since broad native vegetation types were readily discerned. The distribution and density of annual forage were strikingly apparent on the bajadas north of Phoenix because the forage was at a peak (or near-peak) stage of foliage development at the time of the Apollo 9 overflight. Although the timing of the Apollo 9 flight was nearly optimum for discerning the annual forage, considerably more information can be determined from sequentially obtained photographs regarding "range readiness" (when the range can be grazed without damaging the plants) and total availability of the forage resource throughout the various environments in the region. Finally, despite the low resolution of the space photographs, they are extremely useful for stratifying the range landscape into units of similar forage production. Subsampling using ground techniques can then provide accurate estimates of the carrying capacity of such rangelands. The potential productivity of much of the wildland area could also be determined from the space photographs by identifying the important landforms, which are conspicuous on the space photographs, and by understanding the nature of the vegetation, soil, and hydrologic potential associated with these landforms.

The potential for geologic mapping on space photographs was explored in a complex area between Roosevelt Reservoir and San Carlos Reservoir. Also, an evaluation of high-altitude photography for use in geologic reconnaissance was carried out in an area covered by Apollo 9 imagery. The two evaluations were then compared. It was found that because of the low resolution, small scale, and lack of stereoscopic coverage of the satellite photography, only gross delineations and classifications of lithologic units could be made. Young volcanics and alluvial material such as valley fill could frequently be identified correctly. Sedimentary units with topographic expression and continuity of bedding features were identified correctly in many places. However, in most instances, lithologic units, as registered on the satellite photo, were too heterogeneous in topographic expression and tone to be adequately mapped. Some of the larger faults, with about 6 mi of linear topographic expression, could be located.

On high-altitude photography, with its higher resolution and larger scale, more structural information could be extracted than from the satellite photography. Delineation and classification of lithologic units was also more accurate on the high-altitude photography, although the increase in lithologic information gained from using the high-altitude photography was not as great as the increase in structural information from the same photography.

The prospects for delineating hydrologic units, determining the distribution of snowpack, and monitoring reservoir levels from an interpretation of space photos were investigated, and very encouraging results were obtained. Hence, hydrologists can anticipate deriving real benefits from space photography when asked to predict water yield from large drainage basins that draw the bulk of their runoff from seasonal snowpacks.

The value of S065 photography for recognizing urban and surburban features was noted. Applications for recreation planning and water-supply forecasting were discussed. The potential for monitoring changes in land-use patterns was also investigated using data from the Mesa agricultural test site.

SELECTED LITERATURE

AMSBURY, D. L. 1969. Geologic Evaluation of Apollo 6 Photographs (Frames AS6–2–1416 to 1464) of the Ft. Worth-Dallas, Texas, Area. In J. L. Kaltenback, Science Report on the 70-mm Photography of the Apollo 6 Mission. Earth Resources Division, Science and Applications Directorate, NASA–MSC.

64

ARIZONA CROP AND LIVESTOCK REPORTING SERVICE. 1969a. Prospective Plantings for 1969. Mimeographed Rept. 2 pp.

ARIZONA CROP AND LIVESTOCK REPORTING SERVICE. 1969b. Annual Crop Summary. Mimeographed Rept. 2 pp.

BENSON, L.; AND R. A. DARROW. 1954. The Trees and Shrubs of the Southwestern Deserts. University of Arizona Press, Tucson. 437 pp.

BRUNNSCHWEILER, D. H. 1957. Seasonal Changes of the Agricultural Pattern; a Study in Comparative Airphoto Interpretation. Photogram. Eng. 23(1):131–139.

CARNEGGIE, D. M.; L. R. PETTINGER; C. M. HAY; AND S. J. DAUS. 1969. Analysis of Earth Resources in the Phoenix, Arizona, Area. *In* R. N. Colwell et al., An Evaluation of Earth Resources Using Apollo 9 Photography. Final Rept. NASA Contract no. NAS–9–9348. Univ. of California, Berkeley.

COLWELL, R. N.; ET AL. 1969. An Evaluation of Earth Resources Using Apollo 9 Photography. Final Rept. NASA Contract no. 9–9348. Univ. of California, Berkeley.

GOODMAN, M. S. 1959. A Technique for the Identification of Farm Crops on Aerial Photographs. Photogram. Eng. 25(1):131–137.

GOODMAN, M. S. 1964. Criteria for the Identification of Types of Farming on Aerial Photographs. Photogram. Eng. 30(6): 984–990.

PETTINGER, L. R.; ET AL. 1969. Analysis of Earth Resources on Sequential High Altitude Multiband Photography. Spec. Rept. Forestry Remote Sensing Laboratory, Univ. of California, Berkeley.

SAYN-WITTGENSTEIN, L. 1967. The Best Season for Aerial Photography. Trans. Ind. Int. Symp. Photo-Interpretation. Paris, 1966.

SCHEPIS, E. L. 1968. Time-Lapse Remote Sensing in Agriculture. Photogram. Eng. 34(11): 1166–1179.

STEINER, D. 1969. Using the Time Dimension for Automated Crop Surveys From Space. Proc. 35th Annual Meeting Amer. Soc. Photogram. Washington, D.C. pp. 286–300.

WILSON, E. D.; R. T. MOORE; AND J. R. COOPER. 1969. Geologic Map of Arizona. Arizona Bureau of Mines and USGS.

(a) Aerial Ektachrome (conventional color film) hand-held Hasselblad photo taken by the Apollo 9 astronauts from an altitude of 130 mi while over the Imperial Valley, California. This film employs a three-dye emulsion for the blue, green, and red portions of the electromagnetic spectrum.

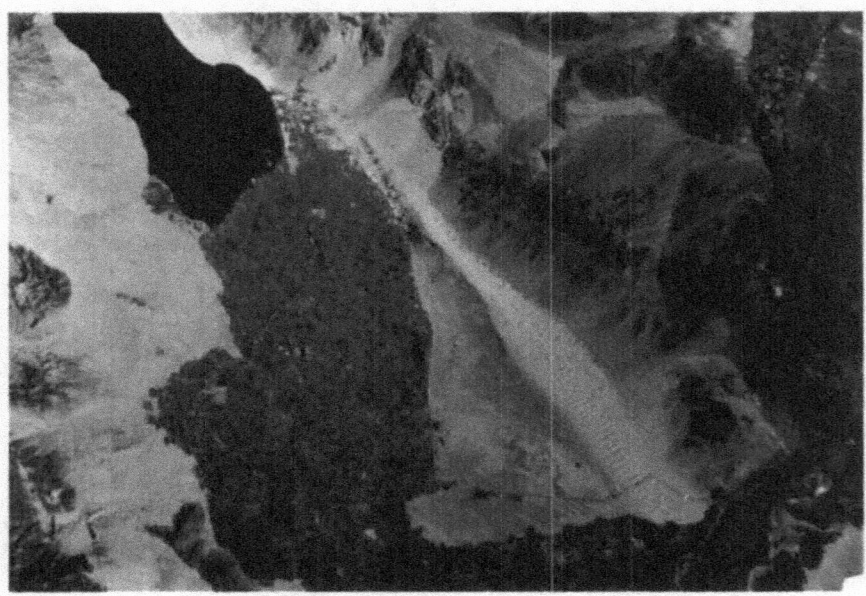

(b) Infrared Ektachrome S065 photo of the Imperial Valley, taken by the Apollo 9 astronauts. This film employs a three-dye emulsion, exposing for the green, red, and near-infrared portions of the electromagnetic spectrum.

Color Plate 17

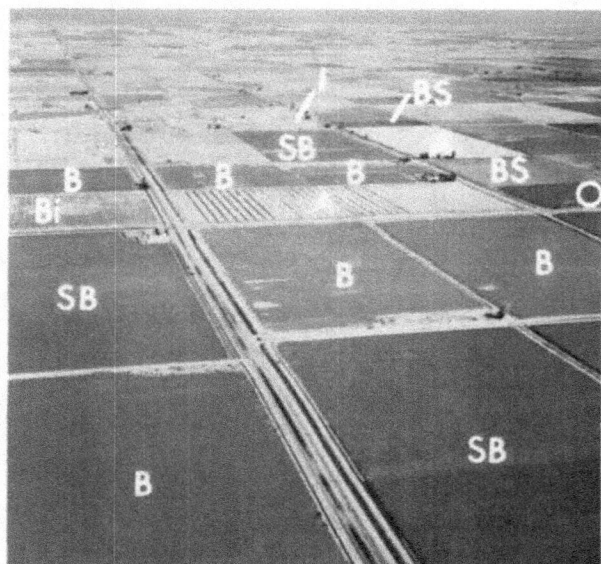

(a) Infrared Ektachrome aerial oblique photo taken over the Imperial Valley on March 9, 1969, at the time of one of the Apollo 9 overflights.

(b) Ten-diameter enlargement of the Infrared Ektachrome S065 photo showing a portion of the Imperial Valley just north of the California-Mexico border. The location of El Centro, Calif., and Mexicali, Mexico, is indicated. The area used for interpretation tests is shown at A, and the area used for interpreter training is shown at B. (Photo scale: 1 in. = 3 mi.)

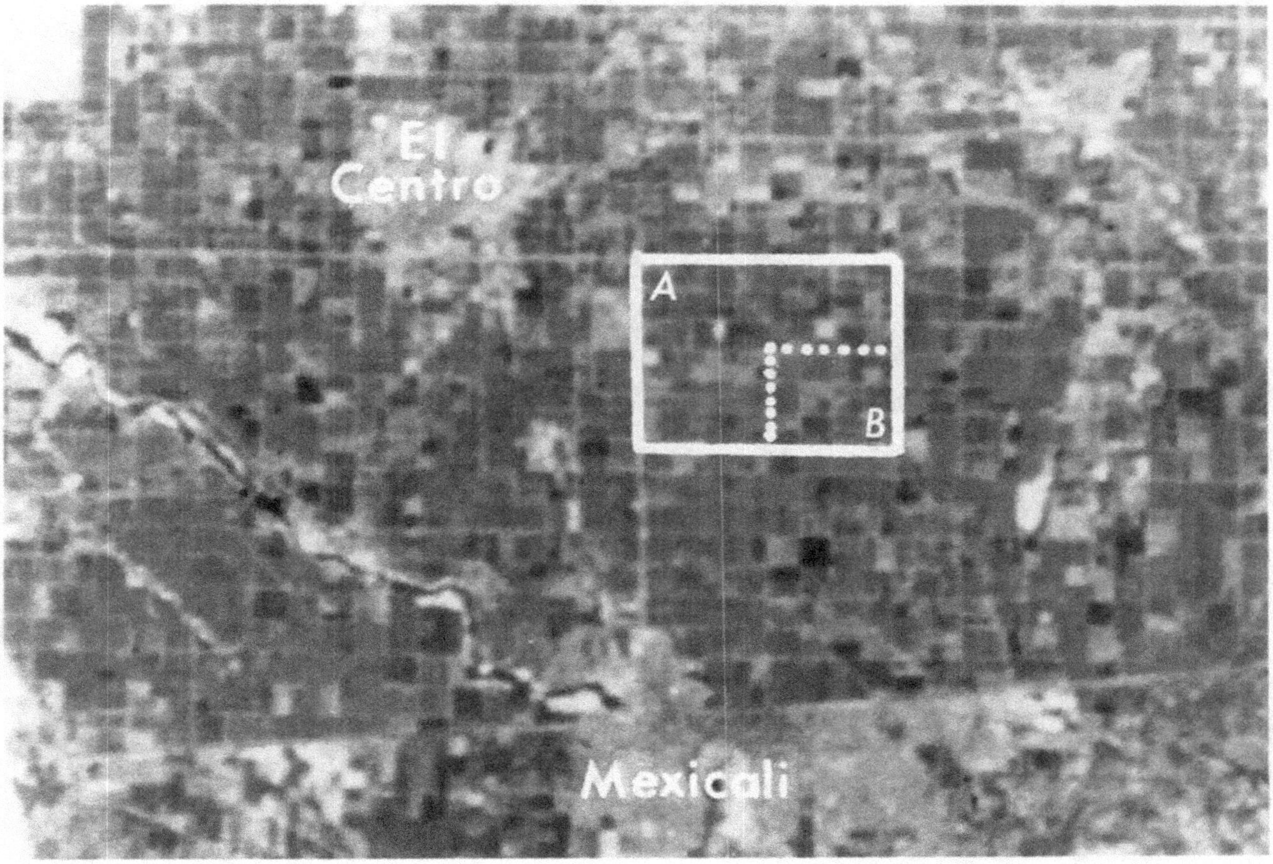

Color Plate 18

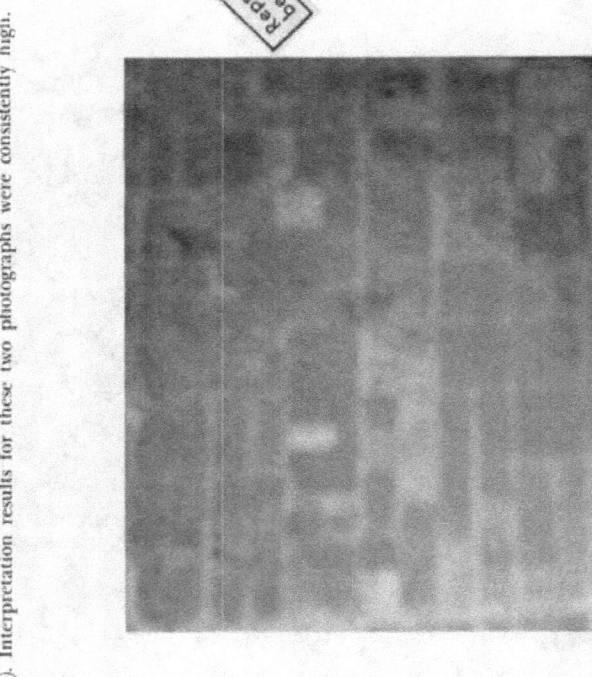

(a) Infrared Ektachrome S065 photo (*left*) and high-altitude aircraft photo (*right*). Interpretation results for these two photographs were consistently high.

(b) These two enhanced images were prepared with the FRSL optical color combiner. The left image was made from the following S065 images and filters: Pan-25A (61 filter) and IR–89B (25 filter). The S065 Pan-58 image was not used because of its poor spatial resolution and low contrast. The right image was prepared using the following high-altitude images and filters: Pan-25A (61 filter), Pan-58 (65A filter), and IR–89B (25 filter).

Color Plate 19

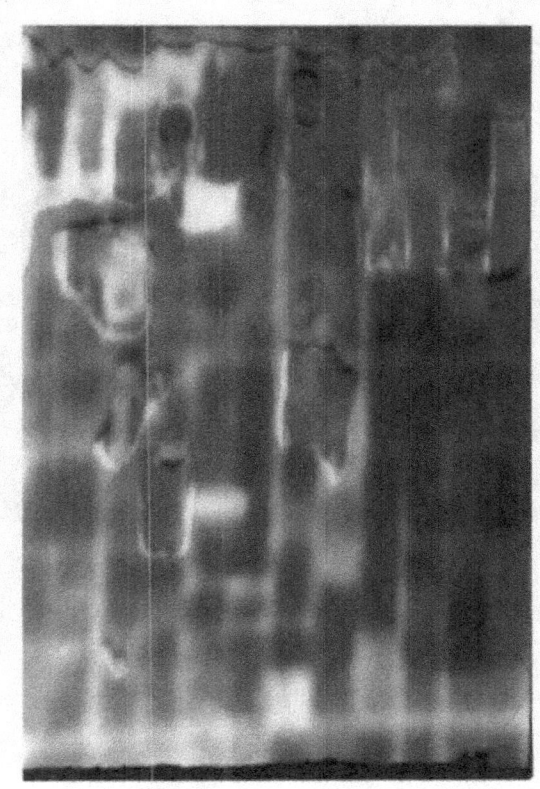

(a) Philco-Ford electronically enhanced images made from S065 imagery (*left*) and high-altitude photography (*right*) using two bands in concert: Pan-25A and IR-89B.

(b) IDECS electronically enhanced images. The image on the left was prepared using three S065 images in concert (Pan-25A, Pan-58, and IR-89B) to highlight bare soil (red). The image on the right was prepared using three high-altitude images in concert (Pan-25A, Pan-58, and IR-89B) to enhance rye (yellow) and barley (orange).

Color Plate 20

Barley, March 1969

Barley, May 1969

Sugar beets, March 1969

Sugar beets, May 1969

Alfalfa (mature), March 1969

Alfalfa (recently cut), May 1969

Color Plate 21. Infrared Ektachrome ground photos, taken during the months indicated, show representative fields of barley, sugar beets, and alfalfa. Monthly changes in crop development aid in the identification of crop type.

Bare soil, March 1969

Cotton, March 1969

Cotton, May 1969

(a) Ektachrome (color) ground photos of bare soil and cotton taken during the months indicated. Use of sequential photography permits these conditions to be identified consistently.

(b) High-altitude aerial photomosaic of the Mesquite Lakes area of the Imperial Valley. The numbers show points from which terrestrial photographs were taken.

Color Plate 22

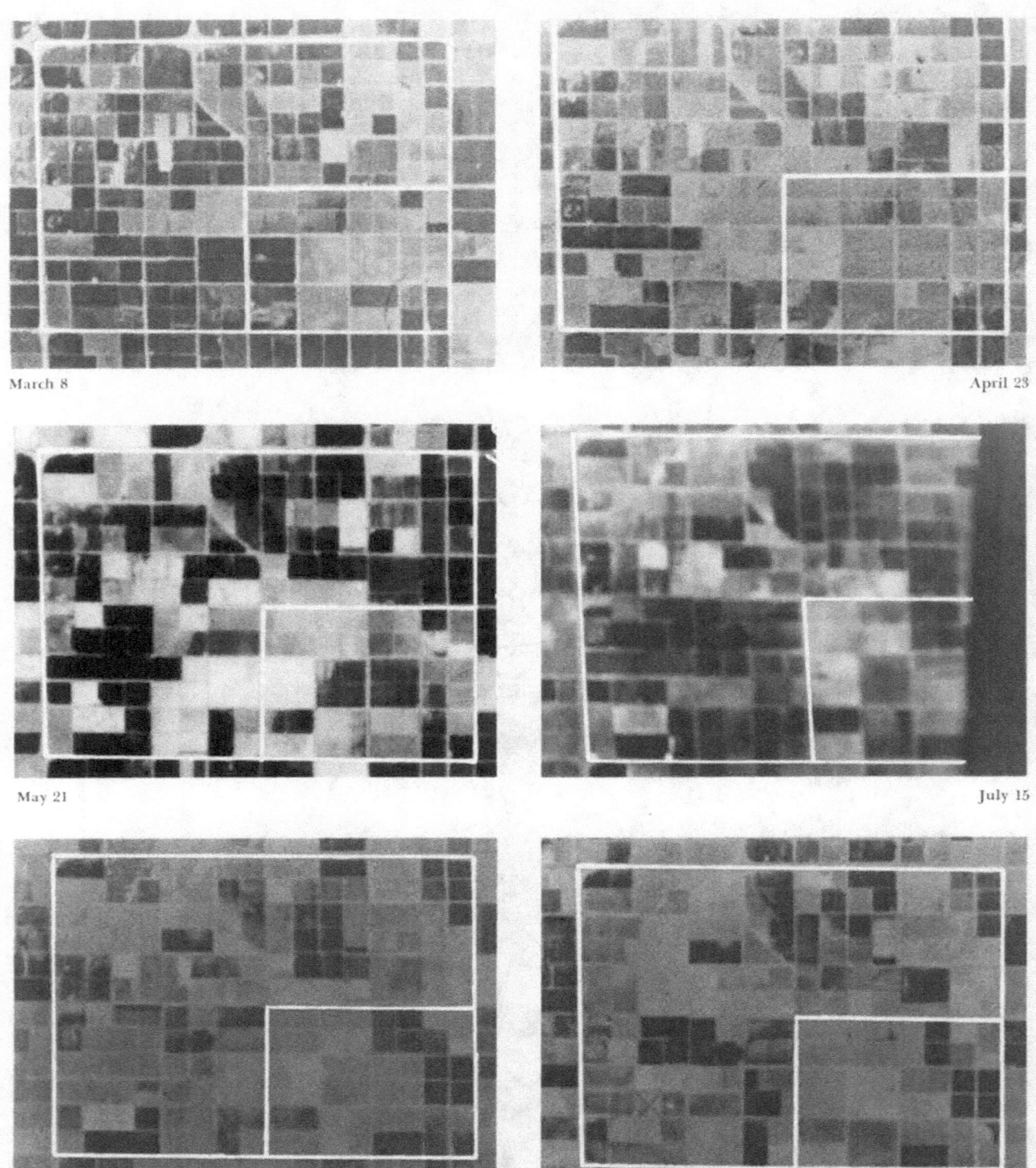

March 8

April 23

May 21

July 15

August 5

September 30

Color Plate 23. Infrared Ektachrome photographs of the Imperial Valley test area (outlined by large white rectangle) taken on the dates indicated during 1969. (Pan-25A appears for May 21 because no Infrared Ektachrome was obtained.) Complete ground truth was collected for every field in this 12-sq-mi area on the day of each photography mission. The area outlined in the lower right corner of each photo is the training area used for the interpretation tests.

(c) Terrestrial photo of barley field severely affected by excessive salinity, taken from point 2 on Color Plate 22 (b).

(d) Terrestrial photo of barley field showing healthy vegetation in reclaimed area, taken from point 3 on Color Plate 22 (b).

(a) Terrestrial photo of an alfalfa field taken from point 1 on Color Plate 22 (b). Sparse areas (brown) in this field indicate that salt accumulations have prevented normal growth of the crop.

(b) Infrared Ektachrome oblique aerial photo taken from point A in Color Plate 22 (b) concurrently with the high-altitude photography in that plate.

Color Plate 24

4

Analysis of Agricultural Resources in the Imperial Valley, California

Randolph R. Thaman and Leslie W. Senger

As noted in the preceding chapter, irrigated agricultural lands are readily differentiated from wildland and urban environments on Earth orbital photography. The study reported in this chapter deals with differentiating individual crops within an agricultural area. An analysis was made of photographs taken by the Apollo 9 astronauts over the Imperial Valley, Calif., and of supporting high-altitude photography. The primary objective was to determine the feasibility of making broad crop inventories on such photography.

The Imperial Valley (Color Plate 17) is in the south-central part of California. It extends from the Salton Sea on the north to the Mexican border on the south and has an areal extent of about 4500 sq mi. With almost a half-million acres of cropped land, the Imperial Valley is the largest single area of irrigated agriculture in the Western Hemisphere, with crop and livestock production grossing more than $230 million a year.

The space photography of the Imperial Valley used for this experiment was taken with the following film-filter combinations:

(1) Infrared Ektachrome film with a Wratten 15 filter (frame AS9–26A–3799A)

(2) Panatomic-X film with a Wratten 58 filter (AS9–26B–3799B) (Pan-58)

(3) Panatomic-X film with a Wratten 25A filter (AS9–26D–3799D) (Pan-25A)

(4) Infrared Aerographic film with a Wratten 89B filter (AS9–26C–3799C) (IR–89B)

To facilitate the acquisition of ground data needed for the evaluation of this photography, complementary vertical and oblique low-altitude photography was obtained over the Imperial Valley by Airview Specialists Corp. of Palo Alto, Calif., almost simultaneously with the overflight by the Apollo 9 spacecraft. This photography shows representative examples of color and tonal variation between different fields due to differences in crop type and field condition. Ground-truth information was acquired on eight dates. The first date coincided with the S065 in March 1969, and the others coincided with the monthly NASA high-altitude flights. Using this information plus the photography obtained during the high-altitude flights, we have sought to determine the extent to which crop inventories can be made with space photography.

The photography of the Imperial Valley obtained during the S065 experiment is of high quality. Color Plate 17(b) and figure 4–1 are examples of this photography obtained by the four-camera multiband system on March 12, 1969, from an altitude of 129 n. mi. Most individual fields are easily discerned. Definite color or tonal differences can be detected between fields; and this factor, as will be shown, facilitates identification of individual crop types and field conditions.

PHOTOINTERPRETATION TESTS

Several photointerpretation tests were designed and administered for the purpose of determining the extent to which agricultural crops in a given test area could be identified. A 12-sq-mi area southeast of El Centro, Calif., was chosen as the test area. Complete ground truth for this area was acquired, coincident with S065, as reported in a document by N. Spansail (1969). At the time of the Apollo 9 overflight, terrestrial photos in both color and Infrared Ektachrome were taken in the test area to document, field by field, the conditions of the crop types. In addition, oblique aerial photos were obtained over selected areas from a low level to document details of field condition and field structure (Color Plate 18(a)). The test area can

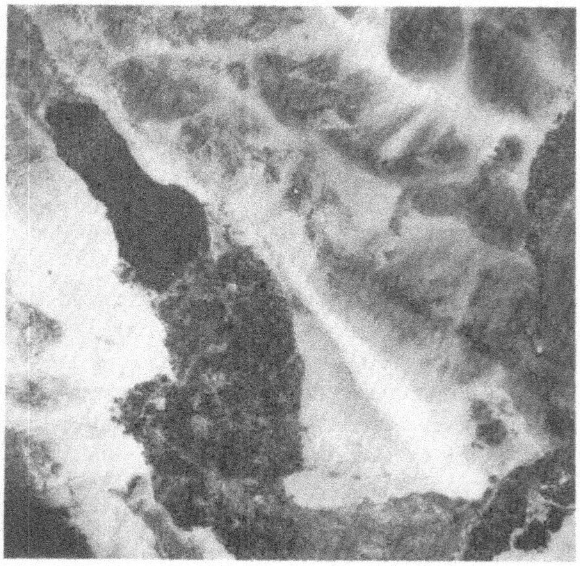

(a) Pan-25A

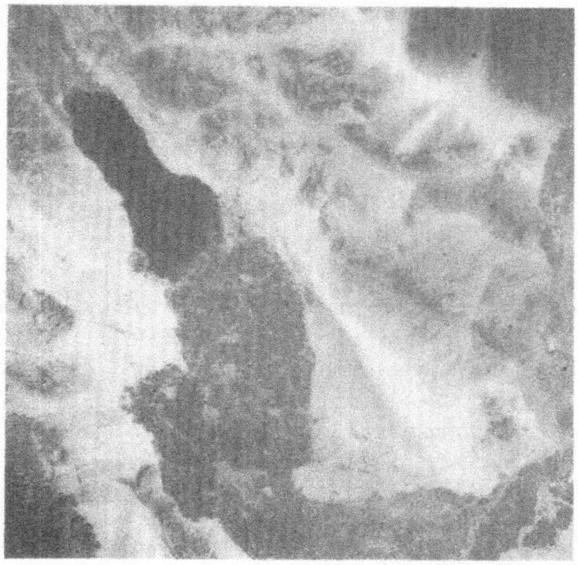

(b) Pan-58

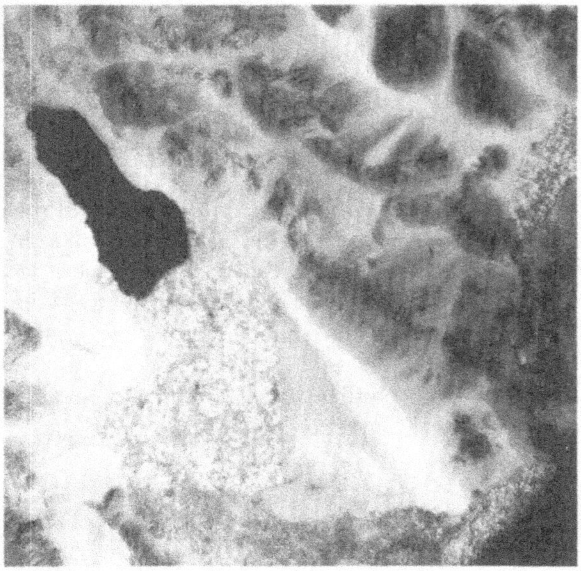

(c) IR-89B

FIGURE 4-1.—Black-and-white multiband S065 photographs of the Imperial Valley taken by the Apollo 9 astronauts. A different film-filter combination was used to obtain each of the photographs so that separate records of reflectance in the red (a), green (b), and near-infrared (c) wavelength bands, respectively, could be obtained.

be seen at A on the 10 × enlargement of the Infrared Ektachrome S065 photograph in Color Plate 18(b). The subarea B, in which the correct identity of crop was given, was used to familiarize the photointerpreters with spectral signatures of known crop types and field conditions.

Image Preparation

The images used for the interpretation test included (1) 20-diameter enlargements of S065 transparencies, (2) NASA high-altitude images enlarged to the same scale, and (3) enhancements made from two or three of the above images. The test images were as follows:

Infrared Ektachrome
Pan-25A
Pan-58
IR–89B
Forestry Remote Sensing Laboratory (FRSL) enhancement
Philco-Ford enhancement

The enhancement systems are described in chapter 2. These images were made into either 3¼-by 4-in. lantern slides or 35-mm slides, which were then projected onto a screen. The interpretation of the photographs was carried out by viewing these enlarged images. The images measured approximately 18 by 13 in. and were viewed from a distance of about 12 ft.

Experimental Design

To obtain quantitative measures of interpretation accuracy, a photointerpretation test was designed and administered. Six categories (crop types and/or field conditions) were interpreted: bare ground versus everything else, alfalfa versus everything else, barley versus everything else, sugar beets versus everything else, rye versus everything else, and overall (all of the first five crop categories interpreted simultaneously). Thirty-six photointerpreters were used in order to (1) obtain statistically significant results, (2) avoid interpreter fatigue, and (3) minimize familiarization with the test area. The interpreters were categorized by photointerpretation experience into three levels of inter-

pretation skill or competence: high, medium, and low.

In the initial phase of the tests, the interpreters were given a test sheet (see fig. 4–2) that shows the fields in the training area B. After studying the ground truth in this training area, the interpreters tried to identify correctly the remaining fields in the entire area A. A sample interpreter sheet appears in figure 4–3.

The symbols of the crop categories used for the test were: BS (bare soil), A (alfalfa), B (barley), SB (sugar beets), and R (rye). Figure 4–4 shows the complete ground truth for the test area acquired on March 12, 1969, coincident with the Apollo 9 overflight.

To reduce bias, interpreters were assigned different categories and image types on a random basis. For both the S065 photography and the high-altitude photography, 36 photointerpretation tests were administered (6 crop categories × 6 image types). Each of the 72 tests was taken by one interpreter from each of three competence levels, making a total of 216 tests.

Each test was corrected to determine (1) the percent *correct identifications* for each category—found by dividing the number of correct responses by the total of correct possibilities and multiplying the result by 100; (2) the percent *omission errors* (i.e., fields that were within the category but which the interpreter failed to identify correctly)—found by subtracting the number of correct responses from the total possible correct, dividing this value by the total possible correct responses, and multi-

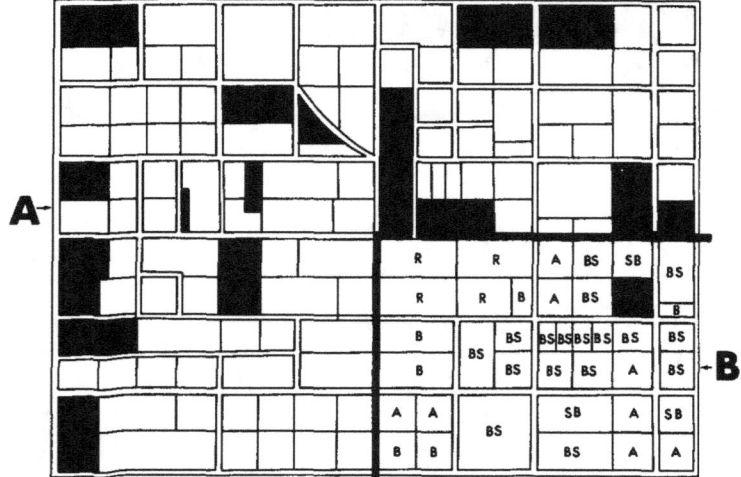

FIGURE 4–2.—Photointerpretation test sheet showing the test area *A* and the ground data for the photointerpretation training area, subarea *B*. (Darkened fields are those with crops other than those designated for interpretation categories.) See the appendix for the crop legend.

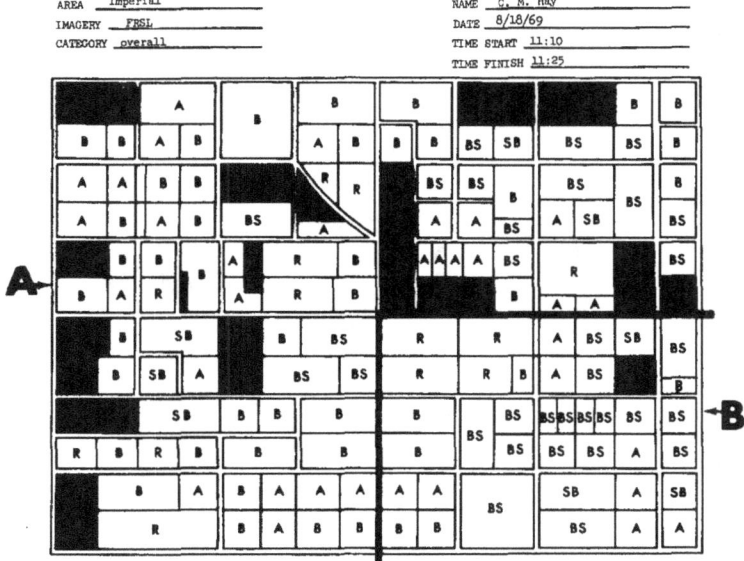

FIGURE 4–3.—A completed test sheet showing how the photointerpreter has used the training area *B* as an aid in identifying crop types and field conditions in the remaining fields of the test area *A*. See the appendix for the crop legend. Total correct, 46; total incorrect, 48; time, 15 min.

FIGURE 4–4.—Total ground data for the 12-sq-mi test area on March 12, 1969, at the time of S065. Subarea *B* is the training area for the interpreters. The lowercase letters (see legend in the appendix) are used for ground data in area *A*. The darkened areas are crops that do not fit into these categories.

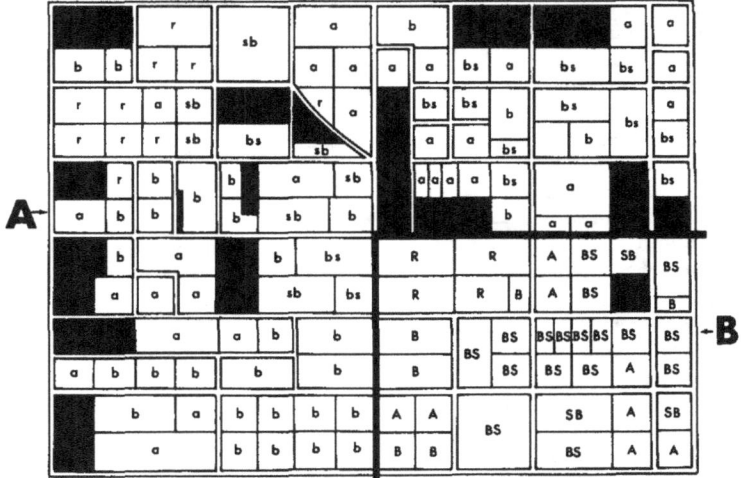

plying the result by 100; (3) the percent *commission errors* (i.e., those fields the interpreter incorrectly identified as belonging to the category under consideration)—found by dividing the total number of responses in that category, minus the correct responses, by the total number of responses and multiplying that result by 100; and (4) the *time*, in minutes, required by the interpreter to complete the test.

Photointerpretation Test Results

In tables 4–1 and 4–2, the results of the interpretation tests are presented. The "rankings"

shown in these tables refer to the rank of that image type based on interpretability for a particular category. Rankings for Apollo 9 images appear in table 4–1, and results for high-altitude images appear in table 4–2. The data show the ranking of a particular image type with respect to the other test images, by crop. It should be emphasized that these rankings are based solely on the raw data. In many cases there may not be significant differences between image types ranked consecutively.

In tables 4–3 to 4–5, each test image is ranked according to (1) percentage correct identifications (all five crop categories plus the overall category

TABLE 4–1.—*Interpretation Results for 6 Types of Apollo 9 Imagery*

Image and criterion	Results by crop type					Overall ranking
	Bare soil	Alfalfa	Barley	Sugar beets	Rye	
Pan-125A:						
Percent correct	75.55	21.87	50.53	50.00	33.33	30.49
Ranking out of 6	4	3	4	3	1	5
Percent commission	16.45	53.91	63.92	87.39	79.09	69.50
Ranking out of 6	5	4	6	3	2	6
Time (min)	7.33	8.33	7.33	9.00	4.33	22.00
Ranking out of 6	6	3	3	4	1	5
Pan-58:						
Percent correct	54.44	16.66	46.23	16.67	20.00	26.59
Ranking out of 6	6	6	5	5	3	6
Percent commission	52.38	62.10	59.07	95.45	94.54	67.01
Ranking out of 6	6	6	5	6	6	5
Time (min)	4.00	8.00	6.00	8.66	10.00	13.00
Ranking out of 6	2	2	2	3	6	1
IR–89B:						
Percent correct	73.33	29.16	39.78	55.55	20.00	37.59
Ranking out of 6	5	1	6	1	3	3
Percent commission	11.11	38.00	34.82	86.89	79.54	62.40
Ranking out of 6	3	1	1	2	3	3
Time (min)	4.00	14.33	5.00	5.67	4.33	17.67
Ranking out of 6	2	6	1	1	1	2
Infrared Ektachrome:						
Percent correct	80.00	19.79	56.98	44.41	23.33	52.48
Ranking out of 6	3	5	3	4	2	1
Percent commission	5.13	50.45	45.74	89.59	57.14	47.55
Ranking out of 6	1	2	3	4	1	1
Time (min)	6.00	9.67	19.33	14.00	7.33	26.33
Ranking out of 6	4	4	5	6	4	6
FRSL enhancement:						
Percent correct	82.22	28.12	67.74	50.00	6.67	42.19
Ranking out of 6	2	2	1	2	6	2
Percent commission	7.51	50.83	51.83	86.82	88.33	57.80
Ranking out of 6	2	3	4	1	5	2
Time (min)	3.33	10.33	7.67	6.33	4.67	18.00
Ranking out of 6	1	5	4	2	3	3
Philco-Ford enhancement:						
Percent correct	84.44	21.87	62.36	11.11	20.00	34.39
Ranking out of 6	1	3	2	6	3	4
Percent commission	11.39	62.02	38.20	95.33	87.90	65.60
Ranking out of 6	4	5	2	5	4	4
Time (min)	6.00	6.67	24.67	9.00	8.33	21.00
Ranking out of 6	4	1	6	4		4

NOTE.—Percent correct, percent commission errors and interpretation time in minutes are shown for each crop or field-condition category. Also, for each category, rankings are given, based on the relative values of the 6 images for identification of that category.

averaged), (2) overall commission errors, and (3) average time per image in minutes.

From these results, it appears that the Infrared Ektachrome imagery gave the best interpretation results.

The Infrared Ektachrome test images (Color

TABLE 4–2.—*Interpretation Results for 6 Types of High-Altitude Imagery (March 1969)*

Image and criterion	Results by crop type					Overall ranking
	Bare soil	Alfalfa	Barely	Sugar beets	Rye	
Pan-25A:						
Percent correct	81.11	52.08	55.26	60.00	23.33	50.00
Ranking out of 6	3	1	3	4	3	1
Percent commission	8.03	59.76	46.01	87.90	81.58	50.00
Ranking out of 6	3	4	3	5	4	1
Time (min)	5.67	11.33	7.00	8.00	8.00	16.00
Ranking out of 6	2	5	2	3	5	4
Pan-58:						
Percent correct	74.76	21.87	34.87	66.66	0.	28.01
Ranking out of 6	5	4	6	2	6	6
Percent commission	33.63	78.88	64.66	81.49	100.00	71.98
Ranking out of 6	5	5	6	2	6	6
Time (min)	7.00	9.33	9.67	8.33	4.33	22.33
Ranking out of 6	5	2	5	4	2	6
IR-89B:						
Percent correct	90.47	15.62	47.20	61.11	43.33	36.87
Ranking out of 6	2	6	4	3	1	5
Percent commission	24.15	84.74	44.81	72.97	80.94	63.12
Ranking out of 6	4	6	2	1	3	5
Time (min)	8.00	10.33	7.00	7.00	9.67	15.67
Ranking out of 6	6	3	2	2	6	3
Infrared Ektachrome:						
Percent correct	97.62	23.96	60.78	38.87	43.33	49.93
Ranking out of 6	1	3	2	5	1	2
Percent commission	6.53	56.82	33.33	87.21	66.01	51.06
Ranking out of 6	2	3	1	4	1	2
Time (min)	6.00	12.00	7.67	6.00	5.67	16.67
Ranking out of 6	4	6	4	1	3	5
FRSL enhancement:						
Percent correct	79.68	31.25	46.38	27.78	10.00	41.48
Ranking out of 6	4	2	5	6	4	3
Percent commission	4.76	41.82	57.86	94.05	77.78	58.51
Ranking out of 6	1	1	5	6	2	3
Time (min)	5.67	5.67	4.67	10.33	3.33	14.33
Ranking out of 6	2	1	1	6	1	2
Philco-Ford enhancement:						
Percent correct	65.55	20.83	74.19	77.78	10.00	37.23
Ranking out of 6	6	5	1	1	4	4
Percent commission	53.29	54.44	54.16	83.90	86.77	62.76
Ranking out of 6	6	2	4	3	5	4
Time (min)	4.67	11.00	11.33	10.00	7.33	14.00
Ranking out of 6	1	4	6	5	4	1

NOTE.—See table 4–1 for explanation of data.

Plate 19(a)) ranked consistently high for almost all categories. The results indicate that subtle differences between crop types can be most easily distinguished on color emulsions. In addition, the haze-cutting ability of Infrared Ektachrome film seems to be a major factor accounting for the con-

sistently good spatial resolution obtained when it is exposed from either aircraft or spacecraft.

The FRSL enhancement also ranked high on many of the overall rankings. (See tables 4–1 to 4–5.) Presumably it ranked high because (1) this image combines spectral returns from more than

TABLE 4–3.—*Ranking of Different Type of Imagery for Percent Correct Identifications (All 6 Categories)*

S065 Apollo imagery [a]		High-altitude imagery	
Imagery	Percent correct	Imagery	Percent correct
Infrared Ektachrome	46.17	Infrared Ektachrome	52.25
FRSL	46.16	Pan-25A	51.96
Pan-25A	43.63	IR–89B	49.10
IR–89B	42.57	Philco-Ford	47.60
Philco-Ford	39.03	FRSL	39.43
Pan-58	30.10	Pan-58	37.70

[a] Statistically significant differences in interpretability of image types were found for these tests (10 percent level of significance). Rankings shown are based solely on raw numerical values.

TABLE 4–4.—*Ranking of Different Types of Imagery for Percent Commission Errors (All 6 Categories)*

S065 Apollo imagery [a]		High-altitude imagery [a]	
Imagery	Percent commission [b]	Imagery	Percent commission
Infrared Ektachrome	49.27	Infrared Ektachrome	50.16
IR–89B	52.13	Pan-25A	55.54
FRSL	57.19	FRSL	55.80
Philco-Ford	60.07	IR–89B	62.45
Pan-25A	61.71	Philco-Ford	65.89
Pan-58	71.76	Pan-58	71.77

[a] See footnote for table 4–3.
[b] Lower percentages indicate more interpretable imagery.

TABLE 4–5.—*Ranking of Different Types of Imagery for Average Time (All 6 Categories)*

S065 Apollo imagery [a]		High-altitude imagery	
Imagery	Minutes	Imagery	Minutes
Pan-58	8.28	FRSL	7.33
FRSL	8.39	Infrared Ektachrome	9.00
IR–89B	8.50	Pan-25A	9.33
Pan-25A	9.72	IR–89B	9.61
Philco-Ford	12.61	Philco-Ford	9.62
Infrared Ektachrome	13.78	Pan-58	10.17

[a] See footnote for table 4–3.

one band, thereby enabling all the information from two or three bands to be seen at once; (2) resolution is quite good; and (3) the color combinations can be varied to allow optimum interpretability. Color Plate 19(*b*) shows the enhancements used for the interpretation tests.

The Pan-25A test results did not rank as high as either the FRSL optical enhancements or the Infrared Ektachrome. The data for each of the individual categories can be seen in tables 4–1 and 4–2. Although the Pan-25A photography (fig. 4–5) does not have the advantage of color differences

72

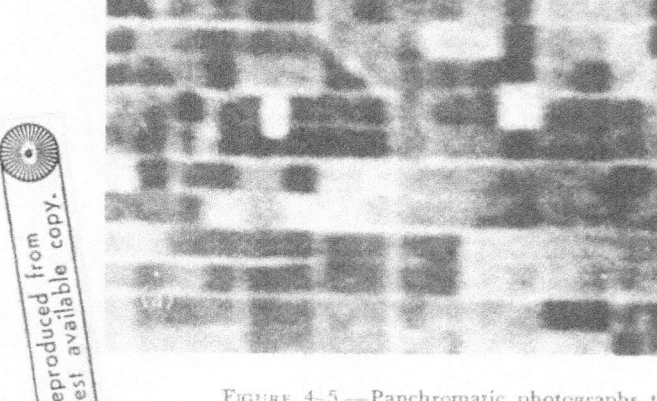

FIGURE 4–5.—Panchromatic photographs taken with a Wratten 25 (red) filter (exposing for only the red wavelengths of the electromagnetic spectrum) (Pan-25A). (*a*) S065 image. (*b*) High-altitude image.

such as those seen on the FRSL enhancements and the Infrared Ektachrome photographs, its higher spatial resolution is a distinct advantage that may explain its high rankings.

The rankings for individual crop categories on IR–89B imagery are somewhat inconsistent. For example, on S065 Apollo imagery (fig. 4–6) IR–89B was definitely best for identifying alfalfa, but on high-altitude imagery it was nearly the worst for identifying this crop. This variation may be a result of having used only three interpreters for each cell. Infrared–89B has very good haze-penetration ability because, for the near-infrared wavelengths to which this film-filter combination

responds, there is very little scattering of radiant energy by atmospheric haze particles. However, the graininess of the film, when enlarged for viewing, detracts somewhat from the advantage afforded by haze penetration.

The Philco-Ford enhancements (Color Plate 20(*a*)) ranked high for some crop categories but for the most part ranked near the bottom. Image enhancements obtained with the Philco-Ford device ranked highest for the identification of bare soil on the S065 imagery. In Color Plate 20(*a*), one can see that the dark-red portion of the enhancement outlines areas of bare soil quite well. Philco-Ford enhancements simplify the work of the

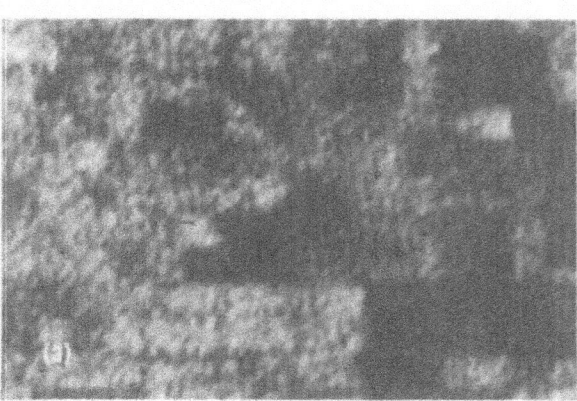
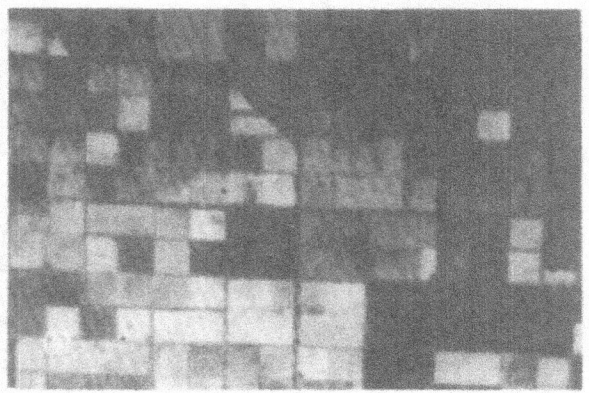

FIGURE 4–6.— Black-and-white infrared photographs taken with a Wratten 89B filter (exposing for only the near-infrared wavelengths of the electromagnetic spectrum) (IR–89B). (*a*) S065 image. (*b*) High-altitude image.

interpreter by using bright colors to enhance subtle tonal variations. The enhancements seen in Color Plate 20(a) were made for the purpose of finding the optimum enhancement for *all* categories on *one* image. Results obtainable with the Philco-Ford device would be improved if the enhancements were prepared with the intent of enhancing for only one category at a time.

Pan-58 photography (fig. 4–7) ranked at the bottom in nearly all of the composite results except for the time criterion on the S065 space photography. The image quality was so poor that many of the interpreters examined it very quickly. Pan-58 definitely shows the least promise for remote sensing from orbital altitudes, since the tone contrast between significant features is poor, as is spatial resolution, due to the scattering of shorter (green) wavelengths by the Earth's atmosphere.

Enhancements from the University of Kansas system (image discrimination, enhancement, combination, and sampling (IDECS)) were not included in the photointerpretation tests. A few IDECS enhancements (Color Plate 20(b)) were made to enhance only a single category (e.g., bare soil from everything else). Consequently they did not meet the format specifications (interpretation for each category and overall interpretation) of the tests reported in this chapter. Although not included in the test, IDECS image enhancements show definite possibilities for facilitating interpretation by enhancing subtle tonal variations and combining the information of a number of different

spectral bands. At present both IDECS and Philco-Ford electronic enhancement systems are still in the experimental stage.

Comparison of S065 Apollo Imagery and High-Altitude Imagery

The data in tables 4–6 to 4–8 compare photo-interpretation results from S065 Apollo imagery with those from the high-altitude imagery. The comparisons in table 4–6 are based on an aggregate of all categories on all six image types. The comparisons in tables 4–7 and 4–8 are based on the results of the FRSL enhancements and the Infrared Ektachrome images, respectively, which were arbitrarily chosen because they seemed to be the optimum image types on an overall basis. The results indicate that there was very little difference in interpretability of S065 Apollo imagery and high-altitude imagery for the particular features studied in this chapter. These results suggest that tonal differences are somewhat more important than spatial-resolution characteristics for identification of crop types and field patterns.

In the tests that have just been described, the accuracy of interpretations was not exceptionally high. The highest percent-correct rating, taking all categories together, was only 52.25 percent. It is important to realize, however, that all these interpretation tests were carried out using no better than third- or fourth-generation images taken at only one time period during the growing season of

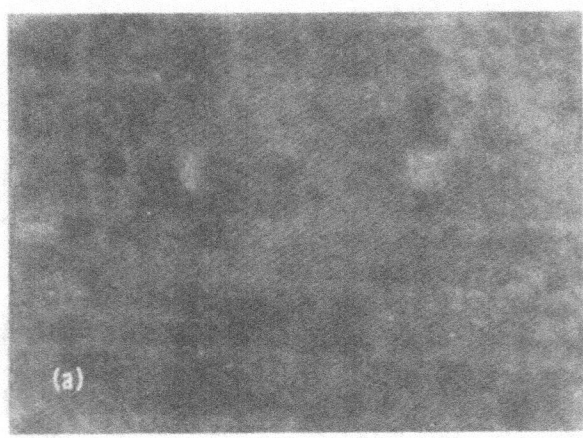

FIGURE 4–7.—Panchromatic photographs taken with a Wratten 58 filter (exposing for only the green wavelengths of the electromagnetic spectrum) (Pan-58). (a) S065 image. (b) High-altitude image.

74

TABLE 4–6.—*Comparisons of S065 and High-Altitude Imagery for an Aggregate of All Categories on All 6 Images*

Percentage correct: S065 Apollo imagery (41.28 percent), high-altitude imagery (46.34 percent)

Standard deviation = 2.79 percent

$t\text{-value} = \dfrac{5.06}{2.79} = 1.81,\ t\!\begin{pmatrix}140\\.10\end{pmatrix} = \pm1.65$ level of significance
Significant difference (10 percent level)

Percentage commission errors: S065 Apollo imagery (58.69 percent), high-altitude imagery (60.27 percent)

Standard deviation = 1.73

$t\text{-value} = \dfrac{-1.58}{1.73} = -0.91$

No significant difference (10 percent level)

Time: average number of minutes per image: S065 Apollo imagery (10.21 min), high-altitude imagery (9.19 min)

Standard deviation = 0.68 min

$t\text{-value} = \dfrac{-1.02}{0.68} = -1.50$

No significant difference (10 percent level)

TABLE 4–7.—*Comparison of FRSL Enhancements From S065 and High-Altitude Images for an Aggregate of All Categories*

Percentage correct: S065 Apollo imagery (46.17 percent), high-altitude imagery (39.47 percent)

Standard deviation = 11.33 percent

$t\text{-value} = \dfrac{-6.73}{11.33} = -.59,\ t\!\begin{pmatrix}34\\10\end{pmatrix} = \pm1.69$

No significant difference (10 percent level)

Percentage commission errors: S065 Apollo imagery (57.19 percent), high-altitude imagery (55.80 percent)

Standard deviation = 10.82 percent

$t\text{-value} = \dfrac{-1.40}{10.82} = -0.12$

No significant difference (10 percent level)

Time: average time per image in minutes: S065 Apollo imagery (8.39 min), high-altitude imagery (7.33 min)

Standard deviation = 1.88 percent

$t\text{-value} = \dfrac{-1.05}{1.88} = -0.55$

No significant difference (10 percent level)

these crops. Much higher percent-correct discriminations could have been made if either sequential photography, or at the very least, a more favorable single date of photography had been utilized. At one given point in time many crops may have almost identical spectral signatures, as was indicated in chapter 3. However, if crops are viewed on other dates, differential phenological changes resulting in different spectral signatures may yield valuable information, enabling interpreters to make more correct identifications.

A prime example of the difficulty faced by an interpreter in identifying crops can be seen in Color Plate 18(a), an aerial oblique (Infrared Ektachrome) photograph taken over the Imperial Valley coincident with the S065 experiment. Even on that large-scale, low-altitude photograph, it was very difficult to differentiate between barley, sugar beets, and mature alfalfa, partly because the photo was not taken at the optimum season of the year in relation to the development of those crops. It can be seen that, although there are definite color differences between fields, some of these are not consistent from crop to crop or even within one crop

type. For example, in Color Plate 18(a) sugar beets (SB) are very hard to differentiate from barley (B) because both crops appear deep red on the Infrared Ektachrome photograph due to their high reflectance in the near-infrared portion of the spectrum. Mature alfalfa exhibits similar spectral characteristics, although it is not represented on this photograph. Even within barley fields, there is considerable variance. For example, on Color Plate 18(a), B represents barley in dense concentrations which show up dark red, whereas Bi represents barley in irregular concentrations which show up pink or even gray on the Infrared Ektachrome photo.

Bare cultivated soil (BS) and uncultivated land (I) show up very clearly as a blue-gray color and are easily differentiated from the surrounding crops. On certain images, some interpreters were able to identify 100 percent of the bare soil fields within the test area. Bare soil was identified correctly more often than any other category due to this distinct tonal or color difference.

In Color Plate 21 (Infrared Ektachrome ground shots), barley and sugar beets can be seen as they

TABLE 4–8.—*Comparison of S065 and High-Altitude Infrared Ektachrome Photographs for an Aggregate of All Categories*

Percentage correct: S065 Apollo imagery (46.17 percent), high-altitude imagery (52.25 percent)

Standard deviation = 8.47 percent

$$t\text{-value} = \frac{-6.08}{8.47} = -0.71, \quad t\binom{34}{10} = \pm 1.69$$

No significant difference (10 percent level)

Percentage commission errors: S065 Apollo imagery (49.27 percent), high-altitude imagery (50.16 percent)

Standard deviation = 9.08 percent

$$t\text{-value} = \frac{0.90}{9.08} = .10$$

No significant difference (10 percent level)

Time: average interpretation time required per image: S065 Apollo imagery (13.78 min), high-altitude imagery (9.00 min)

Standard deviation = 2.70 min

$$t\text{-value} = \frac{4.77}{2.70} = 1.76$$

Significant difference (10 percent level)

appeared in March 1969. At that time of year, both barley and sugar beets reflect a great amount of near-infrared energy giving them a red color as can be seen even more clearly in Color Plate 18(*a*).

When the terrestrial photographs of the same fields taken in May 1969 (Color Plate 21) are compared with those taken in March 1969, it becomes apparent that a marked difference in the spectral signature of barley has occurred, but that little change has occurred in the spectral signature of sugar beets. Barley has matured and appears white on the Infrared Ektachrome photograph because of high reflectance in both the visible and near-infrared portions of the electromagnetic spectrum. Sugar beets, on the other hand, still appear red. Thus, barley and sugar beets, which were very difficult to differentiate in March, can easily be differentiated in May.

The crop that exhibits the most variable signature on all film-filter combinations is alfalfa. At any given time, alfalfa can be found in any number of growth stages, ranging from recently planted through recently cut (7 to 12 cuttings of alfalfa occur yearly in the Imperial Valley). Color Plate

21 shows fields of recently cut and mature alfalfa. Because of its variable signature, alfalfa may be confused at one time or another with almost any crop or field condition; recently planted or freshly cut alfalfa is often confused with bare soil, and mature alfalfa is very hard to differentiate from sugar beets and barley in March photography.

Again, sequential photography could be useful for the identification of alfalfa because the signature of this crop is influenced by repeated cycles ranging from high infrared reflectance to low infrared reflectance throughout the year.

Although bare soil was identified correctly more often than any of the other categories in this experiment, it was sometimes confused with lettuce fields that had been recently disked, and with young cotton and young or recently cut alfalfa. Differentiating between these crops can be facilitated if sequential imagery is used. Color Plate 22(*a*) shows a field containing recently plowed bare soil and a field recently planted to cotton. Both pictures were taken during March 1969. In fields of young cotton, as is true in almost all recently planted fields, little vegetation is seen from an aerial view, making it very difficult to differentiate such fields from bare soil.

The same cotton field, as it appeared 2 months later in May 1969, also is shown in Color Plate 22(*a*). At this time, cotton has matured considerably and, although this is an Ektachrome photograph, there would be markedly greater reflectance of near-infrared energy from the leaves of the cotton plants; thus by using sequential imagery it would also be possible to differentiate crops like lettuce and cotton from bare soil.

MULTIDATE PHOTOINTERPRETATION TESTS

Use of Sequential Photography

The interpretation tests described above have shown that correct identifications for all crops using one date (March 12, 1969) rarely exceeded 50 percent. A major aim of this study was to determine whether the use of sequential photography taken on a monthly basis could indeed improve the accuracy of crop identifications for the Imperial Valley area. An additional objective was to expand the identification capability of sequential photography with a view to predicting what types of crops would occupy the various fields once the crops in them had been harvested. To facilitate these investigations, a model was generated that used crop cycles of fields and associated phenological changes of the vegetation.

High-altitude multiband photography was obtained through the NASA Earth Resources Survey Program on a monthly basis from March to November 1969. Coincident with each flight, complete ground truth for a 12-sq-mi area of the Imperial Valley was obtained to evaluate this model. On the basis of results from the previously discussed interpretation tests, using only March photography, it was decided that Infrared Ektachrome photographs probably would give the most consistent results, especially since FRSL optical enhancements were not available for all months. Although such enhancements ranked high in previous tests, they did not lend themselves to inclusion in this test. The method used for evaluating the model used in this part of our study entailed the administration of photointerpretation tests using the photography from optimum months. The following sections describe the experimental model and give the results of these tests.

A Dynamic Approach to Crop Identification

The model proposed in this section is based on observed crop cycles and associated phenological characteristics as recorded in the Imperial Valley. Without such a model, which includes both criteria, the identification of crops from aerial photos would require an essentially static approach. In contrast, a dynamic approach was taken in this study in that both the horizontal aspects (phenological characteristics of vegetation) and the vertical aspects (crop cycles of fields) were considered. In this way a method was developed for making crop identifications and for predicting the conditions likely to exist in each field at specific future dates.

The model was constructed from several basic elements that can be graphically displayed in the form of a crop calendar. (See fig. 4–9.) The first element is a documentation of the crops grown in an area and their yearly cycles. This information is necessary to determine what crops are to be examined and at what times of the year they occur. The next element is derived from an examination of the crops on an individual basis to determine the rotational cycle over the course of a year. (For

example, lettuce fields may go from lettuce to bare soil to cotton or from bare soil to 70 percent cotton and 30 percent sorghum.) This element yields information on the relative proportions of crops which might replace a given crop over a period of time during the growing season. It is important in this phase to determine the critical months during which a field changes from bare soil to a particular crop, and from that crop to still another crop or back to bare soil. The third element is based on the determination of the months that are optimum for photographic discrimination of particular crops. This can be accomplished by a knowledge of the phenological characteristics of a crop (e.g., mature barley that has turned golden in color) and its corresponding photographic appearance as governed by its reflectance characteristics. An attempt was also made to determine, for each crop type, at what time of year a unique tone signature (or the signature offering least confusion with crops having similar reflectance characteristics) occurred. This was done by examining existing photography.

In summary, the procedure for the identification of crops currently present and for predicting future land use was:

(1) Try to identify a crop during the month when it can be optimally discriminated on the photography; then as a check, use associated critical bare soil months for that particular crop (before, after, or before *and* after the cycle of the crop as necessary).

(2) If a crop does not have a useful appearance in *any* single month, identify it during the month when the fewest other crops exist with which it would be confused on the basis of reflectance characteristics. Then, as a check, use critical bare soil months in the same fashion as before.

(3) Use critical bare soil months, or optimum crop-discrimination months, to predict the occurrence of the next crop in the yearlong cycle; appropriate critical bare soil months or optimum crop-discrimination months can then be used as a check for the prediction.

Questions concerning the validity of a model generally revolve around the assumptions that are made during its inception rather than during its eventual application, although the manner in which data are secured and applied is also important. For this study, four assumptions were made, the first three of which appear justifiable from empirical knowledge of crop characteristics. The assumptions were that (1) fields generally follow a regular sequence of crop rotation; (2) some crops have unique spectral signatures during at least 1 month of their growth cycle; (3) there is probably a month when a field has a spectral response similar to bare soil, i.e., between plantings of successive crops on the same field; and (4) the proportions of various crops in an area remain relatively stable unless there is an economic reason for changing the basic pattern. The fourth assumption is based on knowledge of agricultural practices in the United States and other countries. Efforts are often directed at improving crop yields and/or farming techniques; but crops grown are normally fairly stabilized, and totally foreign crops are seldom introduced. This is attributable to several factors, including crop ecology (environmental needs of particular crops), traditional consumer preferences, knowledge of farming practices associated with particular crops, risks perceived by the farmer in changing to another crop, market orientation of the agricultural economy, and local needs.

Experiment Used in Testing the Model

The value of any model depends on how well it describes reality in a specified area. The area of this study was the Imperial Valley. Ground-truth data collected over a 9-month period were used to test the model for identification of alfalfa, barley, and sugar beets and to predict and verify the occurrence of cotton. An attempt was also made to identify rye on the basis of elimination. The following is a description of how the ground-truth data and photography were studied to evaluate the experimental model for specific crops.

The major crops grown in the Imperial Valley from March to November 1969 were lettuce, barley, alfalfa, rye, sugar beets, melons, cotton, and sorghum. All these crops and others were studied on the basis of duration of their presence, chronological pattern of their appearance, frequency of their occurrence, and spectral signatures at various times of the year.

To determine the optimum dates for the identification of given crops and to prepare a crop calendar, ground-truth acquisition was carried out on a monthly basis from March through November, in conjunction with the monthly high-altitude photography. Color Plate 18(*b*) shows the 12-sq-mi area for which complete monthly ground-truth data were acquired. It will be noted that this is the same area as had been studied in detail on the Apollo 9 photography, as reported in an earlier

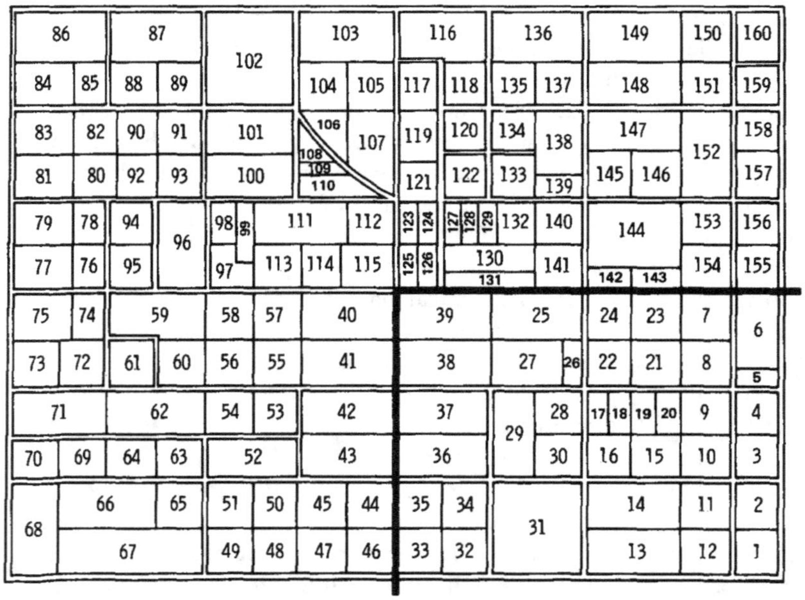

FIGURE 4–8.—Map of the Imperial Valley test site prepared from the March 12, 1969, high-altitude photography. The area contains 160 numbered fields that were visited at the time of each photography mission so that crop type and condition could be recorded. Table 4–9 contains monthly records for each field mapped above.

section of this chapter. Figure 4–8 is a map of the same area prepared from the March 12, 1969, high-altitude photography. The fields were numbered from 1 to 160 to facilitate the study of the progression of crop and field changes during the growing season on a field-by-field basis.

The appendix to this chapter shows the symbols used in the comprehensive study of all 160 fields from March through November 1969. Table 4–9

shows the results of this study, indicating what crop or field condition existed during each month. In the right-hand column a summary of changes that occurred in that field can be found. Where there is more than one symbol in any one entry, the numerator denotes the crop type and the denominator denotes the stage of crop development or field condition. For example, C/VY means cotton, very young (seedlings).

TABLE 4–9.—*Progression of Crop and Field Changes Imperial Valley, Mar. 12 to Nov. 6, 1969*

Field no.[a]	Mar. 12	Apr. 23	May 21	June 30	July 24	Aug. 30	Sept. 30	Nov. 6	Changes
1	A/M	A/M	A/M	A/M	A	A/C	A/C	A/M	N.C.
2	SB/M	SB/M	SB/M	SB/M	BS	BS	BS	A/VY	SB–BS–A
3	BS	BS	C/VY	C/Y	C/Y	C/M	C/M	C/M	BS–C
4	BS	BS	C/VY	C/Y	C/Y	C/M	C/M	C/M	BS–C
5	B/M	B/M	B/D	BS	BS	BS/W	BS/W	BS	B–BS
6	BS	BS	BS	BS	BS/W	BS/W	BS/W	BS/W	N.C.
7	SB/M	SB/M	SB/M	SB/M	BS	BS	BS	BS	SB–BS
8	L/D	BS	S/Y	S/M	S/M	S/C	BS	BS	L–BS–S–BS
9	BS	BS	C/VY	C/Y	C/Y	C/M	C/M	C/M	BS–C
10	BS	BS	C/VY	C/Y	C/Y	C/M	C/M	C/M	BS–C
11	A/C	A/M	A/M	A/C	A/C	A/C	A/C	A/C	N.C.
12	A/M	A/M	A/M	A/C	A/C	A/C	A/C	A/C	N.C.
13	BS	BS	C/VY	C/Y	C/Y	C/M	C/M	C/M	BS–C
14	SB/M	SB/M	SB/M	BS	BS	Ca/VY	Ca/M	Ca	SB–BS–Ca
15	BS	Ca/VY	Ca/M	Ca/H	BS	Ca/VY	Ca/H	A/VY	BS–Ca–BS–Ca–A
16	BS	Ca/VY	Ca/M	Ca/H	BS	Ca/VY	Ca/H	A/VY	BS–Ca–BS–Ca–A
17	BS	Ca/VY	Ca/M	Ca/H	BS	Ca/VY	Ca/H	A/VY	BS–Ca–BS–Ca–A
18	BS	Ca/VY	Ca/M	Ca/H	BS	Ca/VY	Ca/H	A/VY	BS–Ca–BS–Ca–A
19	BS	Ca/VY	Ca/M	Ca/H	BS	Ca/VY	Ca/H	A/VY	BS–Ca–BS–Ca–A
20	BS	Ca/VY	Ca/M	Ca/H	BS	Ca/VY	Ca/H	A/VY	BS–Ca–BS–Ca–A
21	BS	BS	C/VY	C/Y	C/Y	C/M	C/M	C/M	BS–C
22	A/C	A/C	A/C	A/C	A/M	AR/C	AR/C	A/M	N.C.
23	BS	BS	C/VY	C/Y	C/Y	C/M	C/M	C/M	BS–C
24	A/C	A/C	A/C	A/C	A/M	AR/C	AR/C	A/M	N.C.
25	R/M	R/M	R/M	R/W	BS	BS	BS	R/VY	R–BS–R
26	B/M	B/M	B/D	BS	BS	BS	BS	R/VY	B–BS–R
27	R/M	R/M	R/M	R/D	BS	BS	BS	R/VY	R–BS–R
28	BS	Ca/VY	Ca/M	Ca/H	BS	BS	BS	A/VY	BS–Ca–BS–A
29	BS	BS	S/Y	S/M	S/M	BS	BS	A/VY	BS–S–BS–A
30	BS	Ca/VY	Ca/M	Ca/H	BS	BS	BS	A/VY	B*–Ca–BS–A
31	BS	BS	S/Y	S/M	S/M	BS	BS	L/VY	BS–S–BS–L
32	B/M	B/M	B/D	BS	BS	BS	BS	R/VY	B–BS–R
33	B/M	B/M	B/D	BS	BS	BS	BS	R/VY	B–BS–R
34	A/M	A/C	A/M	A/M	A/C	A/C	A/C	A/S	N.C.
35	A/M	A/C	A/M	A/M	A/C	A/C	A/C	A/C	N.C.
36	B/M	B/M	B/D	BS	BS	BS	BS	BS	B–BS
37	B/M	B/M	B/D	BS	BS	BS	BS	BS	B–BS
38	R/M	R/M	R/M	R/D	BS	BS	BS	R/VY	R–BS–R
39	R/M	R/M	R/M	R/D	BS	BS	BS	R/VY	R–BS–R
40	BS	C/VY	C/VY	C/Y	C/Y	C/M	C/M	C/M	BS–C

Footnotes at end of table.

TABLE 4–9 (Cont.)—*Progression of Crop and Field Changes, Imperial Valley, Mar. 12 to Nov. 6, 1969*

Field no.ª	Mar. 12	Apr. 23	May 21	June 30	July 24	Aug. 30	Sept. 30	Nov. 6	Changes
41	BS	C /VY	C /VY	C /Y	C /Y	C /M	C /M	C /M	BS–C
42	B /M	B /M	B /D	BS	BS	BS	BS	SB /Y	B–BS–SB
43	B /M	B /M	B /D	BS	BS	BS	BS	SB /Y	B–BS–SB
44	A /M	A /C	A /M	A /M	A /C	A /C	A /C	A /VY	N.C.
45	A /M	A /M	A /M	A /M	A /C	A /M	A /C	A /VY	N.C.
46	B /M	B /M	B /D	BS	BS	BS	BS	A /VY	B–BS–A
47	A /M	A /M	A /M	A /M	A /C	A /M	A /C	A /Y	N.C.
48	B /M	B /M	B /D	BS	BS	BS	BS	BS	B–BS
49	B /M	B /M	B /D	BS	BS	BS	BS	BS	B–BS
50	B /M	B /M	B /D	BS	BS	BS	BS	BS	B–BS
51	B /M	B /M	B /M	BS	BS	BS	BS	BS	B–BS
52	B /M	B /M	B /D	BS	Ca /Y, BS	BS	BS	SB /Y	B–BS–Ca–BS–SB
53	B /M	B /M	B /D	BS	BS	BS	BS	BS	B–BS
54	A /M	A /M	A /M	A /M	A /C	BS	BS	BS	A–BS
55	BS	BS	C /VY	C /Y	C /Y	C /M	C /M	C /M	BS–C
56	L /D	BS	C /VY	C /Y	C /Y	C /M	C /M	C /M	L–BS–C
57	B /M	B /M	B /D	BS	BS	BS	BS	BS	B–BS
58	L /D	BS	C /VY	C /Y	C /Y	C /M	C /M	C /M	L–BS–C
59	A /M	A /M	A /M	A /C	A /M	A /C	A /M	A /M	N.C.
60	A /C	A /C	A /M	A /C	A /M	A /M	A /M	A /M	N.C.
61	A /M	A /C	A /M	A /C	A /M	A /M	A /M	A /M	N.C.,
62	A /M	A /M	A /M	A /M	A /C	BS	BS	L /VY	A–BS–L
63	B /M	B /M	B /D	BS	Ca /VY	Ca /Y	Ca /M	Ca /M	B–BS–Ca
64	A /M	A /C	A /M	A /C	A /M	BS	BS	L /Y	A–BS–L
65	A /M	A /M	A /M	A /C	A /M	A /C	A /C	A /M	N.C.
66	B /M	B /M	B /D	BS	BS	BS	BS	BS	B–BS
67	A /C	A /C	A /M	A /C	A /M	A /C	A /C	A /M	N.C.
68	L /D, W	BS	BS	BS	BS	Ca /VY	Ca /Y	Ca /M	L–BS–Ca
69	B /M	B /M	B /D	BS	BS	BS	BS	L /VY	B–BS–L
70	A /M	A /C	A /M	A /C	A /M	BS	BS	L /VY	A–BS–L
71	A /C	A /M	A /M	A /C	BS	BS	BS	L /VY	A–BS–L
72	A /M	A /M	A /M	A /C	A /M	BS	BS	L /VY	A–BS–L
73	A /M	A /C	A /M	A /C	A /M	AR /C	AR /C	BS	A–BS
74	B /M	B /M	B /D	BS	BS	BS	BS	Ct /Y	B–BS–Ct
75	A /C	A /C	A /M	A /C	A /C	A /S	AR /S	BS	A–BS
76	G	G	G	G	G	G	G	G	N.C.
77	A /C	A /C	A /M	A /C	A /M	A /C	A /C	BS	A–BS
78	R /M	R /M	R /M	R /D	R /D	BS	BS	R /VY	R–BS–R
79	AR	AR /C	AR /C	AR /C	AR /M	AR /C	AR /C	AR /Y	N.C.
80	R /M	R /M	R /M	R /D	R /D	BS	BS	R /VY	R–BS–R
81	R /M	R /M	R /D	R /D	R /D	R /D	BS	R /Y	R–BS–R
82	R /M	R /M	R /M	R /D	R /D	R /D	BS	R /Y	R–BS–R
83	R /M	R /M	R /M	R /D	R /D	R /D	BS	R /Y	R–BS–R
84	B /M	B /M	B /D	B /H	BS	BS	BS	R /M	B–BS–R
85	B /M	B /M	B /D	B /H	BS	BS	BS	R /M	B–BS–R
86	A /C	A /MS	A /M	A /C	A /M	A /M	A /C	A /M	N.C.
87	R /M	R /M	R /M	S /VY	S /Y	S /C	S /C	BS	R–S–BS
88	R /M	R /M	R /M	S /VY	S /Y	S /C	S /C	R /VY	R–S–R
89	R /M	R /M	R /M	S /VY	S /Y	S /C	S /C	BS	R–S–BS
90	R /M	R /M	R /M	R /D	R /D	R /D	BS	R /Y	R–BS–R
91	SB /M	SB /M	SB /M	SB /M	BS	BS	BS	R /Y	SB–BS–R
92	R /M	R /M	R /M	R /D	R /D	R /D	BS	R /VY	R–BS–R

Footnotes at end of table.

TABLE 4–9 (Cont.)—*Progression of Crop and Field Changes, Imperial Valley, Mar. 12 to Nov. 6, 1969*

Field no.ᵃ	Mar. 12	Apr. 23	May 21	June 30	July 24	Aug. 30	Sept. 30	Nov. 6	Changes
93	SB /M	SB /M	SB /M	SB /M	SB /H	BS	BS	R /VY	SB–BS–R
94	B /M	B /M	B /D	B /H	BS	BS	BS	R /VY	B–BS–R
95	B /M	B /M	B /D	B /H	BS	BS	BS	BS	B–BS
96	B /M	B /M	B /D	B /H	BS	BS	BS	BS	B–BS
97	B /M	B /M	B /D	B /H	BS	BS	BS	BS	B–BS
98	B /M	B /M	B /D	B /H	BS	BS	BS	BS	B–BS
99	Sf	Sf	Sf	Sf	Sf	Sf	Sf	Sf	N.C.
100	BS	BS	C /VY	C /Y	C /M	C /M	C /M	C /M	BS–C
101	P	P	P	P	BS /P	BS	BS	BS	P–BS
102	SB /M	SB /M	SB /M	SB /H	BS	BS	BS	BS	SB–BS
103	A /M	A /M	A /M	A /C	AR /M	AR /M	AR /M	A /M	N.C.
104	A /C	A /C	A /M	A /C	A /M	A /C	A /M	A /C	N.C.
105	A /M	A /M	A /C	A /C	A /M	A /M	A /M	A /C	N.C.
106	R /M	R /M	R /M	R /D	R /D	R /D	R /D	BS	R–BS
107	A /C	A /C	A /M	A /M	A /M	AR /M	AR /M	A /M	N.C.
108	Sf /Y	Sf /Y	Sf /H	Sf /D	Sf /D	BS	BS	SB /Y	Sf–BS–SB
109	O /M	O /M	O /H	BS	BS	BS	BS	SB /Y	O–BS–SB
110	SB /M	SB /M	SB /M	SB /M	SB /H	BS	BS	O /Y	SB–BS–O
111	A /M	A /C	A /M	A /M	A /M	A /C	A /C	A /C	N.C.
112	SB /M	SB /M	SB /M	SB /H	BS	BS	BS	SB /Y	SB–BS–SB
113	SB /M	SB /M	SB /H	BS	BS	BS	BS	BS	SB–BS
114	B /M	B /M	B /D	BS	BS	BS	BS	BS	B–BS
115	B /M	B /M	B /D	B /H	BS	BS	BS	BS	B–BS
116	A /M	A /C	A /C	AR /M	A /M	BS	BS	A /Y	A–BS–A
117	A /C	A /M	A /M	A /M	A /M	BS	BS	SB /Y	A–BS–SB
118	A /M	A /C	A /C	A /M	A /M	A /C	A /C	A /M	N.C.
119	A /C	A /M	A /M	A /M	AR /C	BS	BS	SB /Y	A–BS–SB
120	BS	BS	C /Y	C /Y	C /M	C /M	C /M	C /M	BS–C
121	Sf /Y	Sf /M	Sf /M	P	P	BS	BS	SB /Y	Sf–P–BS–SB
122	A /M	A /M	A /C	A /M	A /C	A /C	A /C	A /M	N.C.
123	P	BS	BS	BS	BS	BS	M /VY	M /M	P–BS–M
124	P	BS	BS	BS	BS	BS	M /VY	M /M	P–BS–M
125	P	BS	BS	BS	BS	BS	M /VY	M /M	P–BS–M
126	P	BS	BS	BS	BS	BS	M /VY	M /M	P–BS–M
127	A /M	A /M	A /C	A /M	AR /C	AR /C	AR /C	AR /C	N.C.
128	A /M	A /M	A /C	A /M	AR /C	AR /C	AR /C	AR /C	N.C.
129	A /M	A /M	A /C	A /M	AR /C	AR /C	AR /C	AR /C	N.C.
130	Sf /M	BS	BS	BS	BS	BS	BS	ML /R	Sf–BS–M(L)
131	Sf /M	BS	BS	BS	BS	BS	BS	ML /R	Sf–BS–M(L)
132	A /M	A /M	A /C	A /M	AR /C	AR /C	AR /C	A /C	N.C.
133	A /M	A /M	A /C	A /M	A /C	A /C	A /C	A /M	N.C.
134	BS	BS	C /VY	C /Y	C /M	C /M	C /M	C /M	BS–C
135	BS	BS	S /VY	S /M	S /M	S /C	BS	BS	BS–S–BS
136	A /M	A /M	A /C	A /M	A /M	A /M	A /C	A /M	N.C.
137	A /M	A /C	A /C	A /M	A /M	A /M	A /C	A /M	N.C.
138	B /M	B /M	B /D	BS	BS	BS	BS	BS	B–BS
139	BS	BS	C /VY	C /Y	C /M	C /M	C /M	C /M	BS–C
140	BS	BS	S /Y	S /M	S /M	S /C	BS	BS	BS–S–BS
141	B /M	B /M	B /D	BS	BS	BS	BS	BS	B–BS
142	A /M	A /M	A /C	A /M	A /C	BS	BS	L /Y	A–BS–L
143	A /M	A /M	A /C	A /M	A /C	BS	BS	BS	A–BS
144	A /C	A /C	A /M	A /C	AR /M	A /M	A /C	A /M	N.C.
145	L /D	BS	S /Y	S /M	S /M	S /C	BS	BS	L–BS–S–BS
146	B /M	B /M	B /D	BS	BS	BS	BS	BS	B–BS

TABLE 4-9 (Cont.)—*Progression of Crop and Field Changes, Imperial Valley, Mar. 12 to Nov. 6, 1969*

Field no.[a]	Mar. 12	Apr. 23	May 21	June 30	July 24	Aug. 30	Sept. 30	Nov. 6	Changes
147	BS	Wm /Y	Wm /M	Wm /H	BS	BS /W	BS	O /Y	BS–Wm–O
148	BS	Ca /VY	Ca /M	Ca /M	Ca /H	BS	BS	A /VY	BS–Ca–BS–A
149	O /M	O /M	O /M	BS	BS	BS	BS	R /VY	O–BS–R
150	A /M	A /C	A /C	A /M	A /M	A /M	A /C	A /M	N.C.
151	BS	BS	S /Y	S /M	S /M	S /C	BS	L /VY	BS–S–BS–L
152	BS	Ca /Y	Ca /M	Ca /M	Ca /H	BS	BS	A /VY	BS–Ca–BS–A
153	L /D	BS	SG /Y	SG /M	SG /C	BS	BS	A /VY	L–BS–SG–BS–A
154	L /D	BS	SG /Y	SG /M	SG /C	BS	BS	A /VY	L–BS–SG–BS–A
155	O /M	O /M	O /M	O /H	BS	BS	BS	A /VY	O–BS–A
156	BS	Ca /Y	Ca /M	Ca /M	Ca /H	BS	BS	A /VY	BS–Ca–BS–A
157	BS	Ca /Y	Ca /M	Ca /M	Ca /H	BS	BS	A /VY	BS–Ca–BS–A
158	A /M	A /M	A /C	A /M	A /M	A /C	A /C	A /M	N.C.
159	A /M	A /M	A /C	A /M	A /M	A /C	A /C	A /M	N.C.
160	A /M	A /M	A /C	A /M	A /M	A /C	A /C	A /M	N.C.

NOTE.—See the appendix to this chapter for definitions of symbols.

[a] Fields 1 through 39 constitute the training area.

TABLE 4-10.—*Progression of Crops and Field Changes for Fields Occupied in March by Alfalfa*

Crop or condition	Number (and percent) of occupied alfalfa fields, by month								
	March	April	May	June	July	August	September	October	November
BS							12 (25)	12 (25)	5 (11)
A	48 (100)	[a]*48 (100)*		48 (100)	48 (100)	48 (100)	36 (75)	36 (75)	36 (75)
L									5 (11)
SB									2 (3)

NOTE.—The data for the optimal month for discrimination of the crop are shown in italic. See the appendix to this chapter for definitions of crop symbols.

[a] The 48 alfalfa fields under consideration are 30 percent of the total 160 fields.

TABLE 4-11.—*Progression of Crops and Field Changes for Fields Occupied in March by Barley*

Crop or condition	Number (and percent) of occupied barley fields, by month								
	March	April	May	June	July	August	September	October	November
BS				28 (94)	30 (100)	30 (100)	29 (97)	29 (97)	17 (58)
A									1 (3)
B	30 (100)	30 (100)	[a]*30 (100)*	2 (6)					
Ca							1 (3)	1 (3)	1 (3)
L									1 (3)
R									6 (20)
SB									3 (10)
Ct									1 (3)

NOTE.—The data for the optimal month for discrimination of the crop are shown in italic. See the appendix to this chapter for definitions of crop symbols.

[a] The 30 barley fields under consideration are 19 percent of the total 160 fields.

TABLE 4–12.—*Progression of Crops and Field Changes for Fields Occupied in March by Sugar Beets*

Crop or condition	Number (and percent) of occupied sugar-beet fields, by month								
	March	April	May	June	July	August	September	October	November
BS				7 (78)	8 (89)	9 (100)	8 (89)	8 (89)	4 (44)
A									1 (11)
Ca							1 (11)	1 (11)	
R									2 (22)
SB	9 (100)	9 (100)	*9 (100)*	2 (22)	1 (11)				1 (11)
O									1 (11)

NOTE.—The data for the optimal month for discrimination of the crop are shown in italic. See the appendix to this chapter for definitions of crop symbols.

ᵃ The 9 sugar-beet fields under consideration are 6 percent of the total 160 fields.

TABLE 4–13.—*Progression of Crops and Field Changes for Fields Occupied in March by Bare Soil*

Crop or condition	Number (and percent) of occupied bare soil fields, by month								
	March	April	May	June	July	August	September	October	November
BS	35 (100)	*19 (54)*		2 (6)	2 (6)	10 (29)	17 (49)	10 (29)	3 (8)
A									11 (30)
C		16 (46)		16 (46)	16 (46)	16 (46)	16 (46)	16 (46)	16 (46)
Ca				11 (30)	11 (30)	3 (7)		7 (20)	1 (3)
S				6 (18)	6 (18)	6 (18)	2 (5)	2 (5)	
L									3 (9)
O									1 (3)

NOTE.—The data for optimal month for discrimination of the crop are shown in italic. See the appendix to this chapter for definitions of crop symbols.

ᵃ The 35 bare soil fields under consideration are 22 percent of the total 160 fields.

In this table a number of characteristic chronological sequences of the appearance and disappearance of crops can be recognized as part of the general system of crop rotation (e.g., lettuce—bare soil—cotton—bare soil). The data for four important crops and bare soil are shown in tables 4–10 to 4–14. The figures in italic in each table show the estimated optimum time for discrimination plus the time when 100 percent of that crop or field condition is in the given classification. For example, in April, which seems to be an optimal month for the identification of alfalfa, all the fields of alfalfa (48) were chosen as the datum base. Then, looking both to earlier and later months, the changes that these fields underwent were tabulated. In each entry for each month and crop there are two important figures.

The first figure represents the number of fields of the crop which were, at each point in time, actually occupied by that crop (or field condition) as listed in the left-hand column of the table. This gives the number of fields of a certain crop that undergo a change and also gives information as to what that change is.

The other figure is the previously described number converted into the percent of the original fields under consideration. The footnote to the optimal month gives the number of fields in the category under consideration during that month and the percent these fields are of the total number of fields in the test area.

Using the data compiled in these tables, the crop calendar in figure 4–9 was prepared. This crop calendar gives the interpreter a general idea of the time of year during which a given crop can be found and at what time that crop can best be differentiated from the other crops on the basis of phenological characteristics and concomitant spectral signatures on high-altitude photography. From knowledge of harvesting and planting times, it is

TABLE 4–14.—*Progression of Crops and Field Changes for Fields Occupied in April by Cotton*

Crop or condition	Number (and percent) of occupied cotton fields, by month								
	March	April	May	June	July	August	September	October	November
BS	16 (100)								
C		16 (100)	16 (100)	16 (100)	16 (100)	16 (100)	16 (100)	*ᵃ16 (100)*	16 (100)

NOTE.—The data for the optimal month for discrimination of the crop are shown in italic. See the appendix to this chapter for definitions of crop symbols.

ᵃ The 16 cotton fields under consideration are 10 percent of the total 160 fields.

also possible to predict times of the year when certain fields will contain bare soil. Bare soil is easily identified on the photography, and the existence of fields of bare soil at critical times, either before or after the occurrence of a crop, can be useful for the verification of assumptions as to the identity in a given field.

Through use of the phenological characteristics of the different crops and the available photography, optimum dates for photographic discrimination of the four major crops were determined: Alfalfa, April 23, 1969; barley, May 21, 1969;

sugar beets, May 21, 1969; and cotton, August 5, 1969.

It is very important to determine (1) when each of the different crops is harvested and when the field turns to bare soil or (2) when bare soil, at a certain time, may indicate what crop will be present in the future. It may be possible to use the chronological appearance of bare soil as a major criterion for crop identification and, at second best, this criterion will be useful as a cross reference for substantiating previous assumptions as to crop identification. Pertinent data for the crops under consideration are as follows:

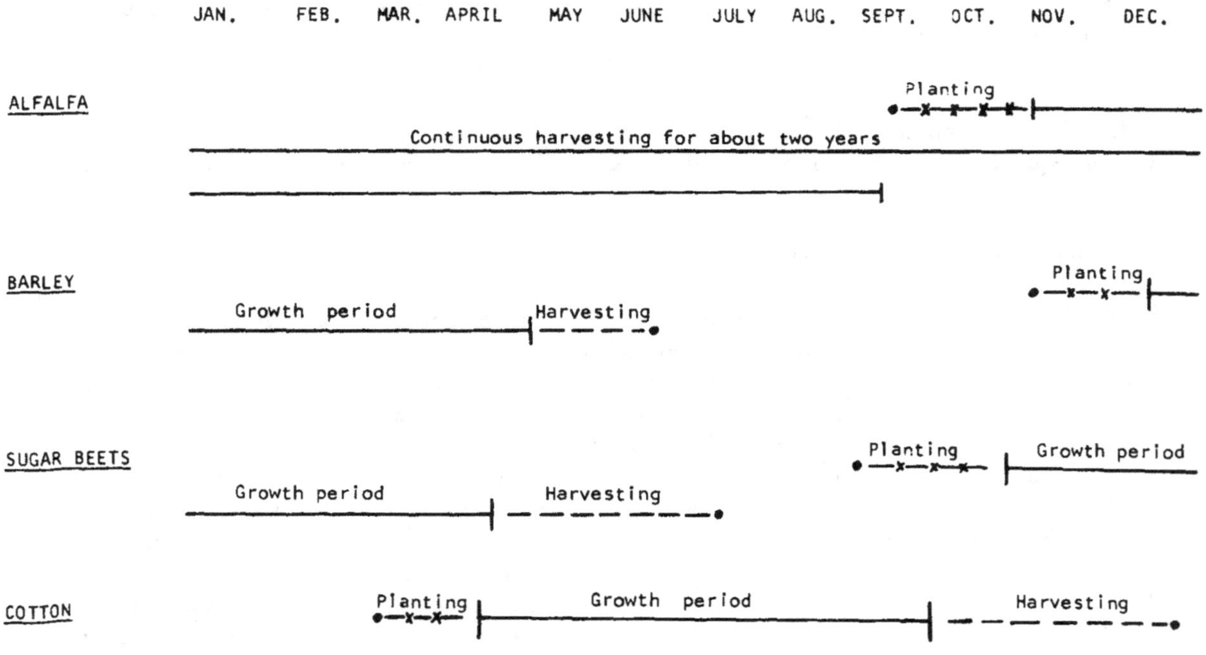

FIGURE 4–9.—Crop calendar indicating cultivation patterns and duration of each crop for four major crop types in the Imperial Valley test area.

Crop	Period When Fields Contain Bare Soil
Barley	July-August
Sugar Beets	June-July
Cotton	January-April
Alfalfa	Usually continuous cover, may be bare soil from September to November before planting (2-year cycles)

Test Procedures

As previously stated, the model was to be tested for four crops (barley, sugar beets, alfalfa, and cotton) using the photographs of the test area that appear in Color Plate 23. Rye and sorghum, also important crops in the area, were to be considered only on the basis of elimination. Rye dries and matures during the summer months while alfalfa stays green; this difference facilitates the differentiation of the two crops. However, because of weather problems and camera malfunctions, the Infrared Ektachrome photography obtained from May through July was not of sufficient quality to permit these differences to be used effectively. There was also no suitable photography during the main portion of the growing season of sorghum.

Each interpreter was asked to study the crop calendar (fig. 4–9) to become familiar with the approximate time and duration of cultivation for each crop in the test area. Four test sheets were provided, two sheets for each test. Each set of pages was numbered from 1 to 160, the numbers corresponding to the field numbers on the map of the test area in figure 4–8. Sample sheets (fields 1 through 80) for each test are shown in figure 4–10. The correct interpretation of each field in the training area was shown so that each interpreter could train himself with respect to the test procedures. The instructions to the interpreters were as follows.

Test 1

(1) Crops to be identified: Barley (B), sugar beets (SB), alfalfa (A), lettuce (L), and rye (R—elimination only)

(2) Photography to be used (1969 dates): March 8, April 23, May 21, July 15, and August 5 (Infrared Ektachrome for all dates except May 21, for which no Infrared Ektachrome is available and Pan-25A is substituted)

(3) The following phenological characteristics should be reviewed for the crops to be identified (color descriptions given for Infrared Ektachrome film):

(a) *Barley* turns golden in color as it matures, dries during April and May, and is harvested by the end of July.

(b) *Sugar beets* do not mature by April or May and tend to have a brighter red color than rye on Infrared Ektachrome photographs taken during that period. Fields of sugar beets are usually covered completely with vegetation (i.e., there are rarely any weak spots or sparse areas as often are found in alfalfa and rye fields), and they are harvested by the end of July.

(c) *Rye* begins to mature as the hot summer months approach, but it does not turn golden to the extent that barley does. By July or August most rye fields appear similar to bare soil on photographs taken in those months.

(d) *Alfalfa* is usually cut every month from February through September or October and will generally have some infrared reflectance throughout the year, ranging from a very limited amount just after cutting to a large amount during periods of maturity, just prior to cutting.

(e) *Lettuce* appears pink to red on Infrared Ektachrome film in March when the crop is maturing. Lettuce fields are harvested and turn to bare soil in April. None of the other four crops just mentioned appears as bare soil during March or April.

(4) The stepwise elimination key in table 4–15 should be followed to carry out the identification of the crops in this test. The photographs should be studied in chronological sequence. Each field should be studied individually and the decision made as to the appropriate category for each field. Photography from successive dates should then be interpreted and eliminations made until the identity of each field has been ascertained. Preliminary decisions should be indicated for each field on the test sheets. The examples in figure 4–10 indicate this procedure for the training area. The training area should be studied carefully to determine how the technique is applied.

(5) Final identifications should be indicated in the column labeled August 5 on the test sheet for all the remaining fields in the test area.

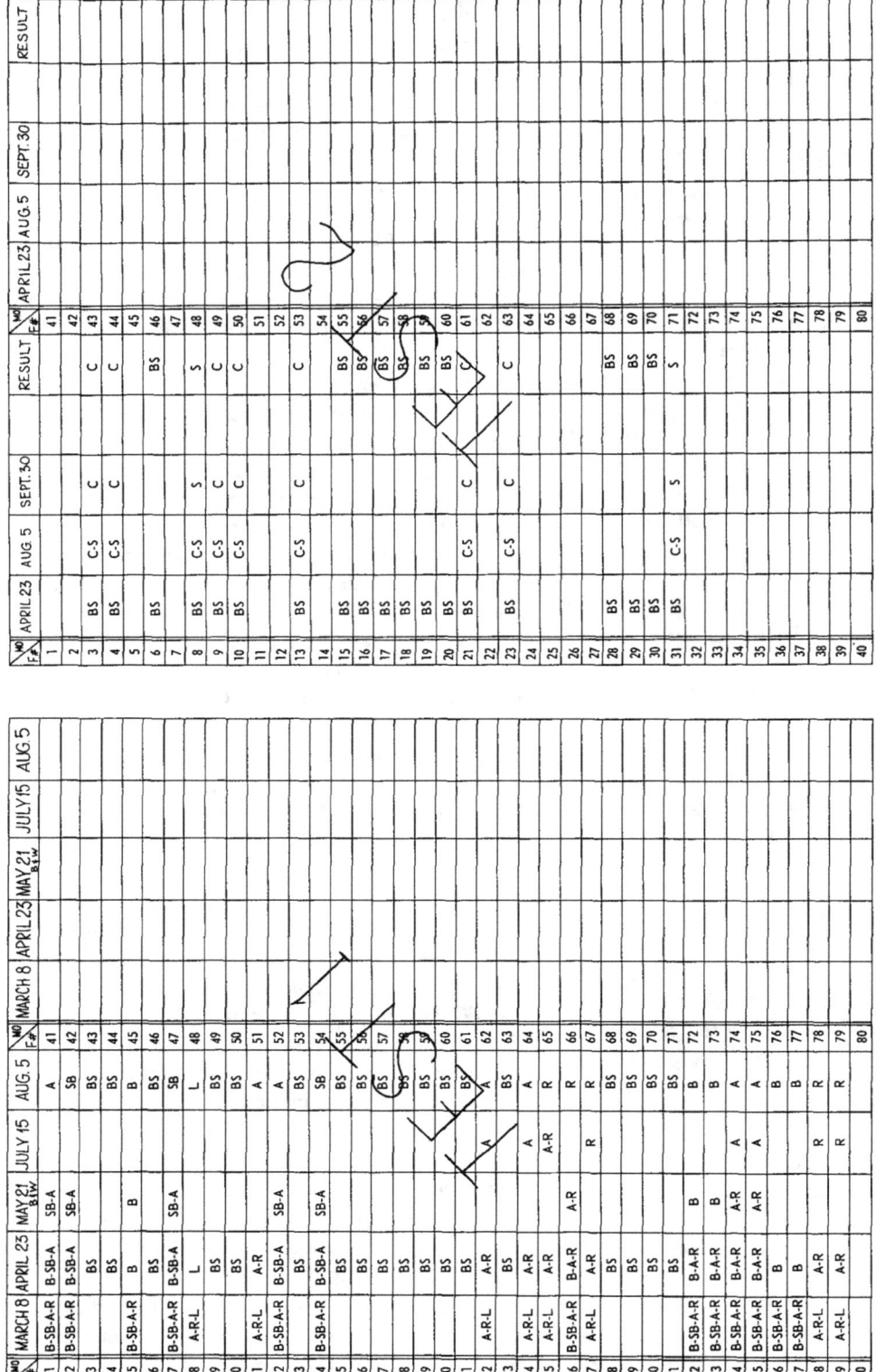

Figure 4–10.—Sample test sheets for interpretation tests 1 and 2 administered for crop identification in the Imperial Valley test area. Interpretation results are indicated for the first 39 fields, which constitute the training area. See text for details of the testing procedure.

TABLE 4–15.—*Eliminative Photointerpretation Key for Test 1*

	Mar. 8		Apr. 23		May 21		July 15		Aug. 5
Color	Crop	Color	Crop	Color	Crop	Color	Crop	Color	Crop
I. Light to dark red	B–SB–A–R	White	B				Go to Aug. 5	Red or pink	*B*
		Bright red, mixed red, or pink and white	B–SB–A	All white; partly white (mottled)	B				*A*
				Gray to black	SB–A	Light to dark red; dark black	*A*	Gray to white	*SB*
		Purplish; sparse dark red or purple	B–R–A	All white; partly white (mottled)	B			Red to pink	*A*
				Gray to black	A–R	Very light (off-white to blue-gray)	A–R	Gray to white	*R*
II. Pink	A–R–L	Gray to blue-gray	L		Go to July 15	Light to dark red; dark black	*A*	Red to pink	*A*
		Any shade of red or pink	R–A			Very light (off-white to blue-gray)	R–A	Gray to white	*R*
III. Gray (no pink or red color at all)	BS (confirm on Apr. 23)	Gray to blue-gray	*BS*; eliminate from further consideration; put BS under Aug. 5						*BS*
		Any shade of red or purple	Go to I or II above						

NOTE.—The identity of each field is determined by selecting the appropriate category on the March 8 photography and using photographs taken on the remaining dates to eliminate all but the final identity of the field. Italicized types are final identifications. See the appendix for definitions of symbols.

Test 2

(1) Crops to be identified: Cotton (C) and sorghum (S)

(2) Photography to be used (1969 dates): April 23, August 5, and September 30

(3) The physiological characteristics upon which the identification is to be based are as follows:

(*a*) Cotton and sorghum plants are too small to have any appreciable infrared reflectance in April and fields containing these crops consequently appear as bare soil.

(*b*) In August, both crop types are in full leaf and have relatively high near-infrared reflectance, but cotton generally has a higher reflectance (resulting in a brighter red image on Infrared Ektachrome film) than sorghum unless the cotton field is sparsely covered.

(*c*) In September, cotton is still in leaf and continues to have high infrared reflectance, whereas sorghum has matured and has lost some of its infrared reflectance (because the leaves are dry and the spikelike inflorescences have matured). In addition, some sorghum fields may have been harvested by this time and bare soil remains in the fields.

(4) The interpretation procedure for this test entails the following steps:

(*a*) Using the April 23, 1969, photography, identify all of the fields that contain bare soil. The sample test sheet for test 2 (fig. 4–10) shows an example of this procedure for the training area.

(*b*) Using only those fields identified in (*a*) as a basis for further identifications, the assumption is next made that these bare soil fields will soon be planted to either cotton (C), sorghum (S), or other (miscellaneous) crops.

(*c*) Using the August 5, 1969, photography, wherever the field has changed to a definite red or pink color the interpreter should call it either cotton (C) or sorghum (S). Such a field should be noted as C–S in the test sheet K (under Aug. 5). If the field looks at all like bare soil it should be eliminated. The field must have a definite red or pink color to be considered as C or S. Those fields that are not C or S should be eliminated from consideration and a dash (—) placed in the appropriate space under the column marked August 5.

(*d*) September 30, 1969, photography is used to differentiate cotton from sorghum. Cotton fields will still appear as a red or light red color (vege-

tated) and sorghum will be light purplish in color (bare soil). On this basis, cotton and sorghum should be distinguished and the final interpretation result placed in the appropriate space in the column under September 30.

Multidate Photointerpretation Test Results

Each of the two tests was administered to four photointerpreters, and each person used the proposed models. The results are presented in table 4–16. In this table, *correct* means the number of fields correctly identified (e.g., 92 for barley), and the percent correct = fields correctly identified divided by total fields in category × 100 (e.g., for barley: 92/104 × 100 = 88.5 percent). *Omission* means the number of fields in the category under consideration that the interpreter failed to identify correctly (e.g., 12 for barley), and percent omission = the number of fields omitted divided by the total fields in category × 100 (e.g., for barley: 12/104 × 100 = 11.5 percent). *Commission* means the number of fields mistaken for categories other than the category under consideration, and percent commission = the number of fields committed divided by total number of fields identified as the category in question × 100 (e.g., for barley: 6/(9 + 6) × 100 = 6.1 percent). A tabulation of all results is presented in table 4–17.

Barley

Results for barley are substantially better than the results of previous tests using only one date of photography. When more than one date was used, it was possible to identify correctly 88.5 percent of the barley fields. Using only one date, March 8, 1969 (for which the photography is of very high quality), it was possible to identify correctly only 60.8 percent of the barley fields.

The percent commission errors using the proposed model are a mere 6.1 percent whereas they were 33.3 percent using only one date.

Results could have been even higher if the interpreters had been more familiar with the testing model. Two fields missed consistently were fields 63 and 141 (fig. 4–8), which looked somewhat like barley in April, but did not exhibit a white tone in May. The interpretation for barley might have been better if the Infrared Ektachrome photography for May had been available to replace the Pan-25A that was used instead. May is the opti-

88

TABLE 4–16.—*Results of Multidate Photointerpretation Tests for Each Crop and Interpreter*

Interpreter, by crop	Total fields	Total correct	Total omission	Total commission
Barley:				
1	26	22	4	2
2	26	24	2	0
3	26	23	3	3
4	26	23	3	1
Grand total	104	92	12	6
Percentage	—	88.5	11.5	6.1
Alfalfa:				
1	43	36	7	5
2	43	33	10	6
3	43	42	1	3
4	43	35	8	4
Grand total	172	147	26	18
Percentage	—	85.0	15.0	11.0
Sugar beets:				
1	6	6	0	2
2	6	6	0	0
3	6	3	3	3
4	6	6	0	8
Grand total	24	21	3	13
Percentage	—	87.5	12.5	38.1
Rye:				
1	11	6	5	3
2	11	7	4	0
3	11	5	6	3
4	11	3	8	2
Grand total	44	21	23	8
Percentage	—	47.7	52.3	27.6
Cotton:				
1	9	9	0	1
2	9	9	0	1
3	9	9	0	1
4	9	9	0	1
Grand total	36	36	0	4
Percentage	—	100	—	10

ᵃ Total fields outside training area.

mum month for the discrimination of barley and a color photograph would yield more information than a black-and-white photograph.

Alfalfa

The results for multidate interpretation of alfalfa show a great improvement over the single-date results. Through use of the proposed model and its eliminative procedure, it was possible to identify correctly an average of 85.0 percent of the alfalfa fields; using only the March 8 photography, an average of only 24.0 percent of the fields were cor-

TABLE 4–17.—*Results of Multidate Interpretation Test for Imperial Valley Test Site, With Results From Tests Using a Single Date, Mar. 8, 1969*

Crop	Percent correct multidate	Percent correct single date	Percent improvement	Percent commission multidate	Percent commission single date	Percent improvement
Barley	88.5	60.8	+24.7	6.1	33.3	−27.2
Alfalfa	85.0	24.0	+61.0	11.0	56.9	−45.9
Sugar beets	87.5	38.9	+48.6	38.1	89.6	−51.6
Rye	47.7	43.3	+4.4	27.6	66.0	−38.4
Cotton [a]	100.0	--------------------		10.0	--------------------	
Sorghum [a]	66.7	--------------------		11.1	--------------------	

NOTE.—High percentages show best performance for correct values, and low percentages are best for commission values.

[a] Because cotton and sorghum were not present in March, a comparison with single-date results is not possible.

rectly identified (with 56.9 percent commission as compared with only 11.0 percent for the multidate test).

These results might have been better if high-quality photography taken during May, June, and July had been available. Most of the omission and commission errors occurred as a result of confusion with rye, which loses considerable infrared reflectance during June and July. If Infrared Ektachrome photography had been available for May, it would have been possible to avoid confusing rye with alfalfa by detecting the presence of bare soil in May.

Sugar Beets

Again, the average correct identifications (87.5 percent for sugar beets) was much higher when using more than one date than when using only the March 8 photography.

Although the commission errors are high, most of them resulted from inability to utilize the July photography to maximum advantage. The results could be improved, especially with regard to the commission errors, if good July photography had been available and if the peculiar growth characteristics of sugar beets could be more objectively described.

Rye

The results for rye were obtained on the basis of elimination (i.e., after identifying other crops), because photography flown during June was not available. During June, rye begins to mature and would be most easily identified at that time.

The percentage correct using the multidate photography, 47.7 percent, is only slightly better than

the 43.3 percent obtained using only March photography, but the percent commission error has been reduced from 66.0 percent to only 27.6 percent by using more than one date.

Cotton

The results using bare soil in April as an indicator of future cropping patterns gave very high reliability for the identification of cotton. No fields were omitted by any of the four interpreters, and the one common commission error was the mistaken identification of sorghum in field 140. This field of sorghum retained high-infrared reflectance and, if October photography had been available, it would have been easily separated from the cotton, which would still have been in leaf. Cotton was not present on March photography, so no comparison was possible for that time.

MONITORING THE EFFECTS OF EXCESSIVE SALINITY

The Imperial Valley is one of the most productive irrigated agricultural areas in the world, but some persons fear that by the year 2000 the farmers in this area will be out of business. Why?

The problem is twofold: An increasing proportion of the cropland in the valley is being adversely affected by (1) a high water table and (2) salt accumulation in the soil. As described by Dr. C. A. Bower, Director of the U.S. Salinity Laboratory at Riverside, Calif.:

The problem is as old as the pharaohs. Every civilization that failed, failed because of the buildup of salt in the soil through irrigation. The civilizations that sprang up along the Nile and also in the

Tigris-Euphrates Valley of Iraq, in India, and in much of the Middle East, all collapsed because they didn't know how to control increasing salinity.

The Imperial Valley is in one of the most arid regions of the United States; its average yearly rainfall is about 3 in. The summers are extremely hot and the relative humidity is low most of the year, with the maximum temperature exceeding 100° F for more than 110 days of the year.

The valley receives its irrigation water from the Colorado River via the 80-mi-long All-American Canal. By the time this water has reached the Imperial Valley, the river and its tributaries have picked up salt from seven States: Wyoming, Colorado, Utah, Nevada, New Mexico, Arizona, and California. Increasing irrigation in the upper part of the Colorado River drainage basin has also caused a considerable rise in the salinity of irrigation water in the Imperial Valley. The plants utilize the water from the soil but do not assimilate most of the dissolved salts, so a considerable amount of soluble salts is left behind after irrigation.

In general, the soils of the Imperial Valley have no real "top soil" or soil profile. The valley is actually a 500 000-acre bowl, filled with a conglomeration of sediments deposited by the Colorado River. In general, there is no sand and gravel to act as a water-bearing stratum. For this reason, the valley is composed of series of aquifers and aquicludes of clay barriers and sand lenses. A stratum of one type does not extend over any considerable area. The valley is almost level throughout and, in the absence of any continuous water-bearing stratum, the salt water does not percolate deeply into the soil. Instead it accumulates and comes to the surface (due to capillary rise) into the root zone of the different crops. The rising water table causes problems when it either drowns the roots of crops or increases the salinity of the soil in the root zone beyond the tolerance level of the crops.

Although excessive salinity is endangering the long-term agricultural potential of the valley, the situation can be improved in various ways if problem areas can be identified.

(1) Drainage canals are constructed by the Imperial Irrigation District to remove irrigation water after it drains off the fields. This water flows into the Salton Sea via the New and Alamo Rivers.

(2) Salt-tolerant crops such as barley and sugar beets can be planted instead of crops that are more adversely affected by salt accumulation.

(3) Drainage tiling can be installed below the root zone to facilitate runoff of saline water that has percolated through the zone of root growth.

(4) Leaching of accumulated salts from the root zone can be accomplished by overirrigating a field or by erecting leach borders and allowing water to stand on a field for a prescribed time.

The first step in solving the problem is to identify fields that need treatment. Although the individual farmers generally know which fields are giving poor yields and which ones may be profitably reclaimed, it may be helpful to use high-altitude aerial photography both to locate problem areas and to plan reclamation projects. It may also be possible to estimate yields of differently affected fields by their appearance on the aerial photography.

The aerial photomosaic in Color Plat 22(b) was taken on March 8, 1969, over the Mesquite Lakes area of the Imperial Valley, south of Brawley, Calif. This is one of the areas most severely hit by excessive salinity. The soils in this area were probably more saline before agricultural development, and the land slopes toward the small river (running from top to bottom in Color Plate 22(b)), thus allowing salt to come to the surface both through lateral movement and capillary rise.

The photographs used in the preparation of the mosaic in Color Plate 22(b) were taken with a HyAc panoramic camera on Infrared Ektachrome film. The healthy vegetation, which was green at the time of photography, appears bright red, and the color of bare soil ranges from dark gray through light brown to white. After close study of the photography, one can see that the effects of excessive salinity show up very clearly, affecting both the soil color and the density of the vegetation cover.

Color Plate 24(a) is a terrestrial photograph taken from point 1 (see photomosaic, Color Plate 22(b)). The alfalfa field has experienced severe damage and decreased yield resulting from excessive salinity. Color Plate 24(a) shows much of the extensive area in which alfalfa has not grown properly. This open or sparse area can be seen clearly on both the mosaic (Color Plate 22(b)) and the oblique Infrared Ektachrome photograph (Color Plate 24(b)), which was taken from point A (on the mosaic in Color Plate 22(b)) at the same time as the high-altitude photographs were obtained.

The terrestrial photographs in Color Plates 24(c) and 24(d) illustrate the great difference between the vigor of vegetation in a field that has been reclaimed and one that is still adversely affected by salinity. Color Plate 24(c) was taken from point 2 (see Color Plate 22(b)). It can be seen clearly on the high-altitude photomosaic that the vegetation growth in this barley field has been severely affected and is far from luxuriant. However, in Color Plate 24(d), taken from point 3, the cover of barley is indeed luxuriant and healthy.

The difference betweeen the fields in Color Plates 24(c) and (d) is very clear on the high-altitude mosaic in Color Plate 22(b). The narrow red strips of vegetation in field 2 are a reaction to the lines of tile that have been laid in the field. Where the tile has been laid the vegetation has reacted favorably, whereas the vegetation between lines of tile has failed to grow.

The lines that appear in field 5 (Color Plate 22(b)) are produced by plant growth as a reaction to leach borders which have been built up to hold water during leaching of the field. As a result of capillary rise, these higher borders have greater salinity than the intervening strips of land; consequently, the resulting vegetation again has a linear appearance on the photography. Field 6 (Color Plate 22(b)) is an example of a field being leached and the leach borders can be seen quite easily.

The very white areas (Color Plate 22(b)) are either those fields which have been abandoned, or crusty accumulations of soluble salts along the river bottom (point 4 on Color Plate 24(b)).

In almost all cases when sparse or weak areas can be detected on the high-altitude photography, these fields have been affected by excessive salinity or a rising water table. High-altitude photography such as that shown in Color Plate 22(b) should greatly facilitate the identification of such problem areas. The small-scale photography, as a result of its broad coverage, gives the interpreter an integrated look at the entire area so that he can locate concentrations of affected fields. This photography might also be used as a conceptual tool to facilitate the timing of broad-scale reclamation programs on a regional basis.

When the vigor of agricultural crops is affected by salinity and other adverse conditions, the yield is reduced. The ability to monitor crop vigor on high-altitude photographs will be of significant interest to those persons who must predict crop yield and identify areas in which there are problems of vigor. Having the synoptic view on very-small-scale photos is a decided advantage over conventional photographs in terms of area seen on one image and ease of handling imagery. Ten photographs at scale 1/20 000 or 20 photographs at scale 1/10 000 would be required to cover the area (25 sq mi) that is covered by the mosaic in Color Plate 22(b), which is made from small portions of HyAc panoramic photography that images a swath of terrain measuring 3 by 40 mi, on each sweep of the lens assembly.

This study of salinity problems in the Imperial Valley is by no means exhaustive, but shows that the effects of soil condition on crops can be detected on small-scale high-altitude photography. With further research, it may be possible to use photography of the type described in this report to improve the management of agricultural land and to identify conditions that affect crop yields.

SUMMARY AND CONCLUSIONS

Although these tests and studies were not exhaustive, they demonstrated that both space photography and high-altitude photographs have great potential applications for resource studies in agricultural areas. The tests indicate that sequential or multidate, high-altitude aerial photography can improve the reliability of crop identifications or inventories. Miscellaneous crops were not included in the photointerpretation model (e.g., lettuce, safflower, onions, carrots, and melons), but the model could be extended to include more crops. With the acquisition of photography during crucial months, i.e., May, June, July, and October, it might be possible to identify a greater variety of crops with a high degree of reliability.

The results of such tests should improve considerably when interpreters become more familiar with the model and with the anomalous changes that often occur within an area such as the Imperial Valley. If it becomes feasible (and the prospects are high) to acquire satellite photography with resolution similar to that of the high-altitude photography used in this study, well-trained interpreters who have worked extensively with very-small-scale photography should be able to achieve better results than the four interpreters who worked for the first time with the model presented in this chapter. Because tonal and color differences are of primary importance for the identification of crop types, the poorer resolution of satellite photography is probably not a major factor inhibiting its utility for studies of this type. Small-scale photography may also be of great value for studies of crop yields and field conditions resulting from such factors as excessive salinity, pests, etc.

The use of satellites in Earth orbit might also eliminate or reduce a serious problem that plagues high-altitude aircraft missions—postponing or canceling photography missions because of cloudiness or other limiting factors. Weather conditions over specific target areas can be monitored from both Earth and space so that a remote-sensing system aboard a spacecraft can collect data whenever the conditions are optimum, but subject, of course, to its being in the right area at the right time. Costly changes in deployment of aircraft can be avoided in this manner and more useful data obtained.

APPENDIX

SYMBOLS USED IN RECORDING THE PROGRESSION OF CROP AND FIELD CHANGES WITHIN THE IMPERIAL VALLEY TEST AREA

	Crop Types		*Field and Crop Conditions*
A	Alfalfa	BS	Bare soil
B	Barley	C	Cut
C	Cotton	D	Dry
Ca	Carrots	H	Harvested
Ct	Cantaloup	M	Mature
L	Lettuce	R	Redivided field
M	Mustard	S	Sparse
O	Onions	VY	Very young (seedling)
P	Pasture	Y	Young (immature)
R	Rye	W	Weeds
S	Sorghum		
SB	Sugarbeets		
Sf	Safflower		
SG	Sudan grass		
Wm	Watermelon		

N.C. No change in crop types or field condition during the observed timespan

SELECTED LITERATURE

ANON. 1968. The Imperial Valley. Windsor Publ., Encino, Calif.

SPANSAIL, N. 1969. Imperial Valley Ground Truth for Apollo 9 Overflight of March, 1969. Rept. no. 2264–7–X. Willow Run Laboratories, Inst. Sci. Tech., Univ. of Michigan, Ann Arbor.

5

A Preliminary Vegetational Resource Inventory and Symbolic-Legend System for the Tucson-Willcox-Fort Huachuca Triangle of Arizona

CHARLES E. POULTON, BARRY J. SCHRUMPF, AND EDMUNDO GARCIA-MOYA
Oregon State University

This chapter is concerned with methods for mapping natural vegetation and with related resource information. The suggested procedures and the mapping example involve the application of cartographic principles and symbolic-legend concepts developed in prior resource-analysis research at Oregon State University.[1] We have found that the same principles and concepts, and many of the procedures, appropriate to resource inventory and analysis from large-scale aerial photography are applicable when space and small-scale high-altitude photographs are the working material.

DESCRIPTION OF THE STUDY AREA

The area studied is about 1200 sq mi in southeast Arizona (fig. 5-1) and includes three large basins. One of them is drained by the Santa Cruz River in the west, and one is drained by the San Pedro River in the eastern half of the triangle. Both rivers flow to the north through the study area. The third basin includes Willcox Playa. Several mountain ranges within the area rise abruptly above extensive sloping bajadas and pediments. Some portions of the area are sparsely dotted by small hills of erosion-resistant material. The flood plain of the Santa Cruz River is 2600 to 3000 ft in elevation. The broad sloping plain, consisting of lower and upper bajadas, rises from

the river to the mountains at 4000 to 5000 ft. The mountains, in turn, rise to an elevation of 9400 ft. The flood plain of the San Pedro River within the study area descends from 4000 to 3400 ft. These broad sloping plains rise to the mountain bases at 5000 ft, and the mountains reach up to 8200 ft. A smaller structural basin lies between the two main rivers and is drained into both of them. It is nearly surrounded by mountains and lies within a 4000- to 5000-ft elevational range.

The dominant climate of the area is semiarid, but humid uplands occur within the semiarid region. The winter precipitation is brought by migratory cyclones. The summer rainfall is of monsoon nature, occurring as ephemeral thunder-and-lightning storms that drench small areas. Although the precipitation is patchy, the entire area will receive precipitation in the course of the summer rainy season from these frequent storms.

Several factors combine to reduce the effectiveness of the summer rainfall: (1) intensity of the rainfall, (2) topography of the mountains and plains, (3) soil surface texture, and (4) vegetational cover. The rain during the summer storms often falls faster than soil infiltration can occur. Sheet floods wash the interfluves of plant and animal debris that would otherwise contribute organic matter to the soil. Loose, fine-textured mineral grains are also removed, leaving behind a gravel- and stone-covered surface. This erosion creates myriad enmeshed runnels that are classified as a parallel drainage pattern (fig. 5-2). The water infiltrates mainly in the stony and/or sandy bottoms of the large channels in the basin drainages. This provides moisture to support large shrubs and

---

[1] Research in the development of new methods for resource analysis supported by the Bureau of Land Management, U.S. Department of the Interior, and in an operational resource inventory and analysis of over 600,000 acres of rangeland for the Oregon State Land Board and Division of State Lands.

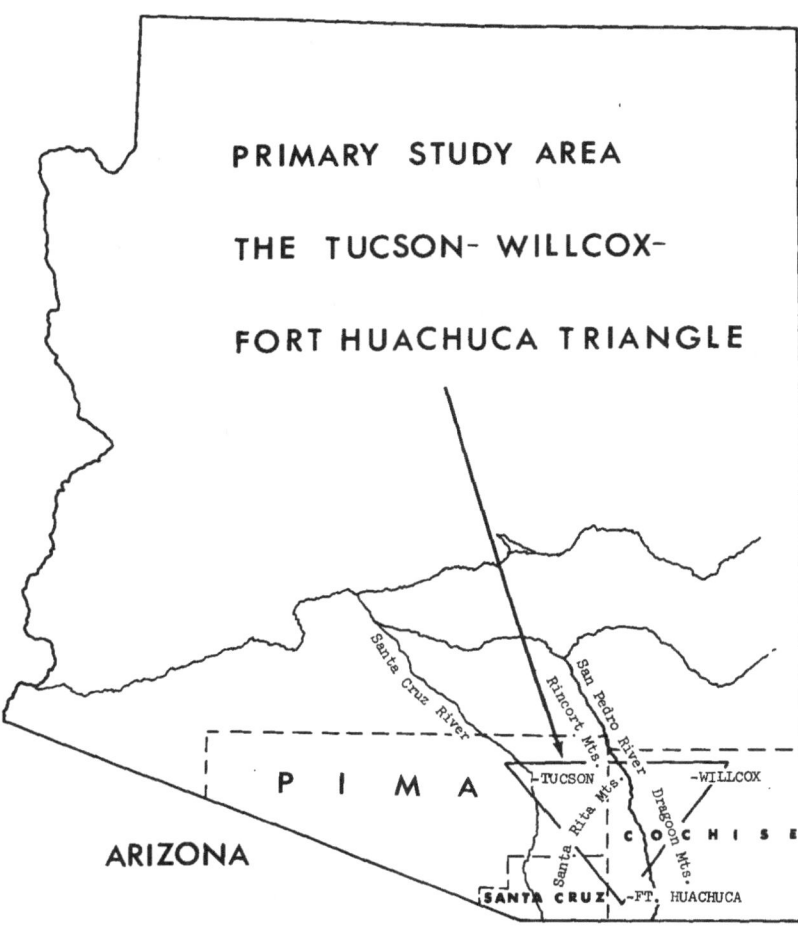

FIGURE 5–1.—Map of Arizona showing the S065 southeastern Arizona test site. Scale = 1:2 534 400.

trees that sometimes have nearly continuous cover (e.g., as in some mesquite bosques). As a result, the most xeric sites occur on the interfluves, the most mesic along the large drainage channels. The runnels present several intermediate environments between these two extremes. These sites occur on the upper and lower bajadas that constitute the sloping plains and the major portion of the desert topography.

The Tucson-Willcox-Fort Huachuca triangle includes part of the eastern fringe of the Sonoran Desert and discontinuous patches of vegetation with a similarity to that of the Chihuahuan Desert. The rest of the test area is occupied by a high elevated basin (4000 to 5000 ft) and its surrounding mountains (8000 to 9000 ft). The former supports a grassland, and the latter are covered by chaparral, juniper-oak woodlands, and/or conifer forests.

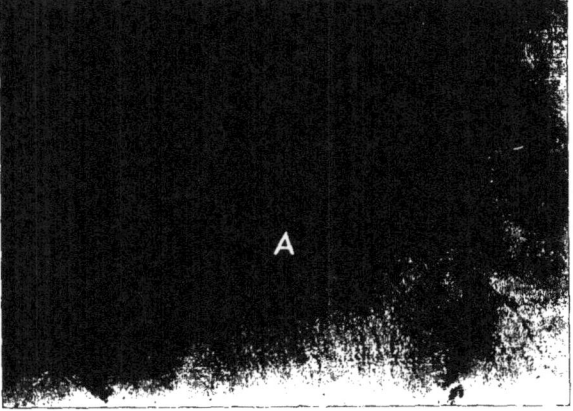

FIGURE 5–2.—High-altitude Panatomic-X, Wratten 58 (Pan-58) image; scale 1:150 000. The enmeshed runnels at A constituting a parallel drainage pattern are a typical erosion pattern in the desert. The accompanying vegetation is desert shrub with little or no herbaceous cover.

Within the grassland area, approximately 70 percent of the annual precipitation falls during the summer. Here, there is sufficient ground cover to hold the water in place while much of it infiltrates the soil. The drainage pattern developing under these vegetational conditions, interacting with soil and geologic materials, is strikingly different. It is classified as dendritic (fig. 5–3).

The forest types lie in zones around the mountains. The upper and lower limits of each zone vary according to aspect. Protected canyons and slopes cause the higher elevational types to extend well below their normal limits, whereas ridges and exposed slopes permit the lower types to extend into higher elevations. This produces an obvious interfingering of vegetational types in many of the foothill and mountain areas.

VEGETATIONAL RESOURCE MAPPING

Remote sensing from space or aircraft platforms produces information of value to the natural resources policymaker, planner, or manager when (1) interpretable information is delineated, identified, and meaningfully symbolized on maps of suitable scale; (2) relevance of the data to resource-management objectives is explained; and (3) the information is abstracted and summarized in a statistical form appropriate to practical information needs. To achieve these conditions, it is important to observe certain cartographic principles and symbolic-legend concepts.

Adherence to three cartographic principles enables one to tolerate easily, without loss of useful information, the residual variability in the way well-trained interpreters delineate vegetational or other resource features:

(1) Separate pure image classes (simple mapping units) wherever cartographically feasible; and when delineating mixtures of image classes (complex mapping units), hold to a maximum of four the number of image subjects comprising the set within a single delineation.

(2) For these complex mapping units, the legend must denote each individual component of the set, *not* an average of the characteristics of the set. Included features, or components comprising less than 10 percent of the delineation, should be ignored in symbolization and avoided in data gathering.

FIGURE 5–3.—High-altitude Pan-58 image; scale 1:150 000. A dendritic pattern has developed in the alluvium at *A* under a semiarid climate. The vegetation at this location is primarily grassland.

(3) The legend should explain each delineation in terms of its factual, present characteristics, *not* in terms of observer interpretations of these characteristics. Interpretations such as ecological climax, productive potential, land-use suitability, etc., are a part of analysis and follow the basic inventory. They should be presented as supplemental overlays, variously colored maps, and/or statistical summaries.

If these principles are adhered to, the low residual disparity in delineation among well-trained interpreters will not cause problems for the user, because each interpreter will have explained his delineations in terms of the individual components of each complex, and accurate facts will have been presented about the resource.

Where native vegetational resources are involved, the approach to legend development is important. The initial objective is to develop a symbolic mapping legend that enables the image analyst to annotate the specific ecosystem or sets of ecosystems that are represented by each unique image. We define "ecosystems" as unique and fundamental ecological units of landscapes having analogous effective environments. Separate examples of an ecosystem may be spatially disjunct across the landscape, but each is identified and characterized by the specific plant community or vegetational association, that occupies the ecosystem area.

Developing a legend requires the acquisition,

analysis, and reduction of ground-truth information by—

(1) Conducting phytosociological studies to identify and describe the ecosystems typical of the landscapes under study

(2) Working out relationships between these vegetation units and their environmental components—soil, physiographic, geologic, and climatic characteristics

(3) Determining the unique image classes that represent the area under study

(4) Relating these image classes to the specific ecosystem, or to the ecosystem sets that each represents

With these steps completed, it is possible to prepare symbolic and descriptive mapping legends by which each image area can be characterized and a large amount of information about the resource communicated to users.

MAPPING LEGENDS FOR SPACE AND AIRCRAFT PHOTOGRAPHY

The natural resource manager has need for almost unimaginable volumes of information as background for his decisions and action programs. His information about the physical resource is nearly all ecologically based, and since range, forest, watershed, and often recreation managers are concerned primarily about vegetational and soil resources, the natural resource manager's initial interest is toward the plant ecology and vegetation-environment relationships evident on the landscapes. To this, he adds facts and understanding relating to the broader human ecosystem—the psychological, sociological, economic, and political environments of man—in reaching his decisions.

To synthesize this complex package of information to a useful point, the manager must be provided with classifications and a way of reducing the data to a point of comprehensibility. This is one of the functions of the ecological legend in resource analysis. The complete package consists of a symbolic legend and a descriptive legend. The former is a kind of shorthand that accomplishes much of the above objective of classification, information, and reduction. In addition, the symbolic legend makes it possible to record tremendous amounts of information in small spaces on maps and in tabular summaries. The descriptive legend, on the other hand, allows the user to rebuild, in its complete form, the detailed information about each symbolized unit.

A Legend Concept

Our research has concentrated heavily on the classification work required to adapt long-established legend concepts to the analytical and mapping requirements of Earth resources imagery. Special attention has been given to the ecology of natural landscapes and to the integrated treatment of land use where man has sharply modified the natural environment and altered uses of the land and continental water resources.

In the use of any remote-sensing system where it is anticipated that ground checking of human or automatic interpretive decisions will be held to the minimum consistent with required accuracy levels, it is important that symbolic and descriptive legends be flexible. It should be possible to update and correct or make additions to the information without reworking the symbolic and descriptive legends. It is also important that the symbolic legend progress from the general, or broad, easily interpretable features, to the more intricate and difficult. With this feature, the annotation of delineation characteristics may be cut off at the point where strong doubt begins to arise as to the accuracy of the statements of the interpreter or image analyst. We have devised and extensively field tested such a system. It is proposed here for consideration and use by other investigators in remote sensing of vegetation and related resources (fig. 5–4).

Fundamentally, the system provides a closed legend using a numeric notation that treats, in the left-hand digits, the broadest and most easily interpretable features in a numerator/denominator symbol format. The numerator treats vegetational, barren-land, water-resource, or land-use features with a logically organized system of descriptors. The denominator provides for environmental descriptors of ecologically and managerially significant features of the Earth's surface. In both cases, features of increasing detail and/or difficulty of interpretation are symbolized progressively as one moves from left to right through the symbol. The system is brief and highly compatible with computerization and data-management needs.

Generalized Form:

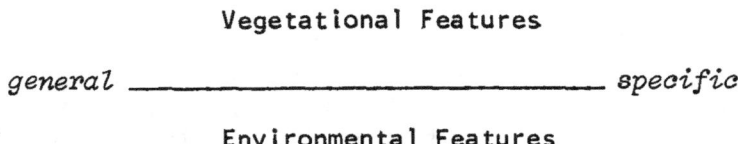

Specific Form:

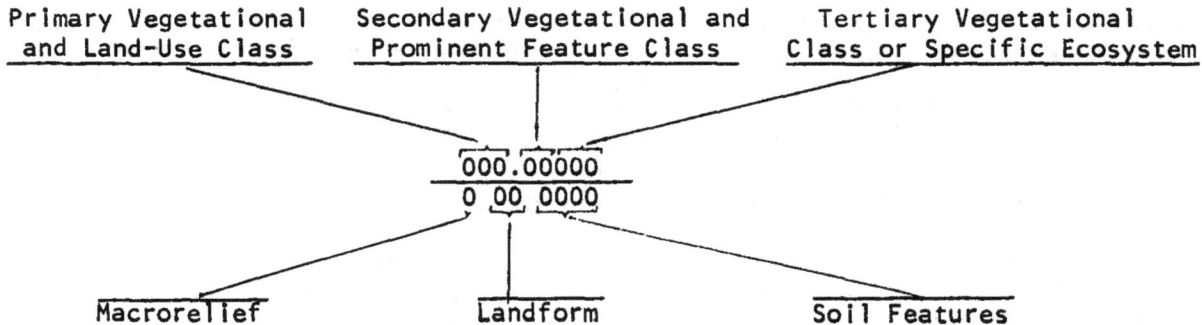

FIGURE 5–4.—An ecosystem legend format for range resource and land-use analysis from space and supporting aircraft imagery.

Physiographic Legend Components

Research in this area is focused on the identification of descriptors that accurately and consistently convey an ecologically meaningful impression of the land surface. To achieve this objective, one must study relationships between vegetation, landforms, and related features, and from this knowledge synthesize and revise the definitions that have been developed by geomorphologists, geologists, and soils people. This work has provided criteria for three sets of descriptors that classify Earth surface characteristics into macrorelief, landform, and microrelief. All three are important to the resource ecologist; and for many practical pur-

poses, resource analysts may well start with the characterization of these features.

Macrorelief

Macrorelief should be taken as the broadest or highest order of generalization in the classification of land surfaces for ecological relevance. It should describe the larger areas that are tied together by similarities in (1) the amount of elevational difference, or relief; (2) the nature and complexity of slopes and abruptness of slope changes; and (3) the complexity of drainage patterns. It is logical, therefore, to set up macrorelief classes that describe natural land-surface conditions ranging from flat and smooth to extremely steep and rugged, and from simple drainage patterns to complex patterns. With these thoughts in mind, we have prepared and widely tested the following set of ecologically useful descriptors of this largest category, or highest hierarchical level, in the classification of land-surface characteristics:

(1) Flatlands
(2) Smoothly sloping, undulating, and rolling lands
(3) Hilly and strongly dissected lands
(4) Mountainous lands

Each is described in appendix 5–A. On high-resolution color and Infrared Ektachrome photography of appropriate scale, many image features are directly related to vegetational characteristics; but as scale and resolution decrease, vegetational interpretations must rely more and more on associated and convergent evidence. A rich ecological experience and fund of knowledge about vegetation-environment relationships is required to identify the criteria from associated and convergent evidence that improve the subject-identification decision. Macrorelief and attendant landforms are two of the most useful kinds of associated evidence in vegetation interpretations. Prestratification into these alternative classes reduces choices in the decision process and tends to increase the accuracy of identification, particularly among less experienced vegetation interpreters (table 5–1).

Space photography taken in 60 percent stereo overlap, as in Apollo 6, is particularly appropriate for mapping. The SO–121 film type and Sun angle at which this photography was taken in the test area are particularly good for the detection of variations in the physiography. In addition, stereo viewing not only gives a good rendition of the relief variations that are important in deciding on

macrorelief classes, but binocular reinforcement also greatly enhances one's interpretive ability in deciding on vegetational features. Macrorelief descriptors are included in the mapping symbols of Color Plate 25.

Landforms

The development of landform classes that are meaningful to the ecologist and vegetation resource manager seems to present a real dilemma. The primary need of these people is for landform classes that are relevant to vegetation development and productivity as well as significant in resource-management decisions. Except for mechanical problems of access and utilization of the resource area, landform features that are relevant to vegetation or to vegetation-soil systems are also the ones relevant to management. Many landform features important to the geomorphologist produce the same effect vegetationally. Thus, if we attempt to use directly the many class names from the literature, we are plagued by synonymy in ecological impact among "separate" landforms.

It appears again that the ecologist and resource manager must improvise their own systems for treating these relevant physical features of the Earth's surface. Vegetation responds strongly to the slope and exposure characteristics of the land surface. Many different landforms, as technically defined, can produce the same slope-exposure environment. The ecologically important point is not the technical landform class, but the angle of incidence of the slope to the Sun's rays and the exposure of the slope to prevailing winds and storms. In addition, the interaction of the slope, the solum, and its substratum are important in determining the soil moisture regime. A suitable system of landform classification should reflect the following eco-

TABLE 5–1.—*Use of Macrorelief and Landform Identification in Vegetational Interpretation*

Macrorelief and landform class	Most likely vegetation or ground feature (legend symbol)[a]
Flatlands:	
Bajadas, fans, or terraces	21.1, 21.2, 21.3, 22.1, 22.2
Bottomlands	22.3, 43.7, 43.81, 43.82, 11.0, 100.0, 92.0
Hilly lands	21.4, 54.0, 65.0
Mountains	54.0, 65.0, 66.0

[a] See app. 5–E through 5–G for description of the symbols.

logically important differences in the land surface:

(1) Bottomlands and lowlands versus uplands

(2) Exposed versus protected slopes

(3) Steepness of slope, length of slope, and position on slope where these features are relevant to vegetational change

(4) Those landform classes that result from strongly contrasting or unique geological influences that are particularly relevant to the ecosystems that are thus delimited by the landform.

It appears that many landforms, as defined by Earth scientists, not only can but should be combined for use in plant sociological research and its related fields of application, one of which is the use of space and high-altitude photography for vegetational resource analysis and land-use planning and management.

Considering these points, the Oregon team has put together a classification of relevant physical features that has worked reasonably well (app. 5–B). This classification has been through many revisions and has been tested from the southern coastal plain to the southwest and northwest, with reasonable success. Only a few of these classes can be used in the interpretation of space photography; but in the setting of multistage ecological analysis of Earth resources, they play a strong role. Again, as with the macrorelief classification, they convey information of value about the landscape and also aid the photointerpreter or image analyst by providing associated and convergent evidence in the image-identification process.

Microrelief

Microrelief refers to the minute inequalities in the land surface within an otherwise homogeneous ecological site. The use of these notations is in more completely characterizing ground-truth locations and the individual ecosystems imaged by some large-scale, high-resolution remote-sensing systems. The microrelief detail is, therefore, omitted from this presentation.

Vegetation and Related Legend Components

Native vegetation, when expressed in terms of the natural plant communities, is the best index to environmental similarity or equivalence, and thus to the inherent productive capacity of the land. Therefore, the two most important ground features

to document in explaining or interpreting images are the vegetational characteristics and the soil surface conditions. Since our space photography studies have been in a rather restricted area, we have prepared a legend for vegetationally related features that is locally adapted, but which could be expanded to fit the region without difficulty. It will also be apparent that some aspects of the legend are continental (or even worldwide) in scope. A decimal symbol system keyed to hierarchical levels of plant community or land-use classification fits all the legend requirements (fig. 5–4).

Primary Vegetational and Land-Use Classes

The numerical descriptors for the primary vegetational and land-use classes are shown in appendix 5–C. The units digit and those to the right of the decimal point (tenths and hundredths) enable the designation of additional resource detail. Eleven classes are included in the set consisting of seven native vegetational categories: one barren lands, one water resources, and two land-use classes. All these have been encountered in the southern Arizona test site except class 70, alpine or arctic tundra, and class 80, vegetation of aquatic environments. An attempt was made to design these classes for worldwide applicability.

Secondary Vegetational or Prominent-Feature Classes

A subordinant breakdown of class 10, barren land, is shown in appendix 5–D. This has been extensively tested and works well. These classes are oriented to landscape ecology and practical management needs. Most of the features are readily identifiable from small-scale photography.

The secondary vegetational legend for primary categories 20 through 60 in the Tucson-Willcox-Fort Huachuca triangle is presented in appendix 5–E. These descriptors occupy the units and the tenths- and hundredths-decimal positions in the numeric symbol. The classes at this level are based on prominent floristic features that are common to sets of similar ecosystems. While the primary classes seem to fit worldwide conditions, it appears that some of the secondary classes may have to be developed separately for broad ecological regions or provinces. In developing this legend, we have examined the work of many well-recognized plant geographers, but most were designed from ground surveys and were not intended for interpretation of vegetational resources from photographic images. It appears, therefore, that operational legends will be a synthesis of many efforts like these, plus qualified judgment about compatibility of the legend with image analysis and interpretation objectives. Some aspects of this legend are still undergoing minor modification but, in general, it is working well for the subject area. The legend has been used to a limited degree in the Phoenix-Mesa test site where the need for some modifications are indicated, particularly in the finer categories. Parallel concepts have been used with success in the northwest.

Secondary legend descriptors for classes 90, water resources (app. 5–F); 100, agricultural land (app. 5–G); and 200, urban and industrial land (app. 5–H), have also been developed and subjected to limited testing. These units are thus combined into a comprehensive legend system with a uniform logic. We believe this should be achieved for successful use of data collected by Earth Resources Technology Satellites and Earth Resources Operation Satellites. Our work on nonvegetation components of the legend duplicates that of other disciplines. It is noted, however, that existing proposals for other numeric legends do not adequately treat the range resources area; and yet other features of these legends can be combined with minor modification, as part of the expanded detail under our latter three primary captions. In setting up these classes, we have been motivated to make the higher levels as compatible as possible with image-interpretation capability from small-scale, moderate- to coarse-resolution imagery.

Uses of the Legend

The legend has been used in three ways: (1) to annotate mapped delineations on space photography, (2) to identify vegetation and related resource subjects on the supporting NASA aircraft photography and on the high-altitude photography, and (3) to identify vegetation quickly at ground-truth observational stations and from small-aircraft overflights.

Vegetation Mapping

As a first step in the analysis of landscapes for vegetational or land-use management, it is usually worthwhile to map the broad macrorelief classes (and if photoscale permits, the landform or relevant feature classes) that are relevant to vegetation growth, to ecological succession, and to man's activity in land-use and resource development.

The refinement of vegetational mapping is dependent on (1) the scale of the photography used, (2) the suitability of the film and filter type for detecting significant vegetational boundaries, and (3) the skill and field experience of the interpreter. When working with space photography, it is rarely possible to map pure plant sociological units. These scales, however, do permit rather refined delineations of vegetations into sets that have characteristics in common and that offer somewhat comparable management, use, and development potentialities.

The approximation of a vegetation map of the test area is shown in Color Plate 25. The legends for the delineation and scientific names of the plants referred to are given in appendix 5–C (Primary Vegetation and Land Use), appendix 5–D (Barren Land), appendix 5–E (Secondary Vegetation), and appendix 5–I (Species List). Because of the scale problem, most mapping units represent complexes of different vegetational associations or plant communities. Where they are highly similar and closely held together by common vegetational characteristics, they are described as a single legend entry, but as multiple entries where the kinds of vegetation within the delineation are strongly contrasting. In many instances, these vegetational contrasts are not detectable from space photography without taking advantage of the near-infrared reflectance of green vegetation and multi-seasonal imagery of multistage photography.

Multistage Resource Analysis

The progressive refinement of descriptors made possible by this legend concept is ideally suited to multistage subsampling with various scales of photography. This application is illustrated in Color Plate 26. Operational procedures were simulated by using selected frames from the various scales of photography flown along limited flight lines we had designated for other purposes. The procedure calls for progressive mapping, interpretation, and image identification at legend-recognition levels appropriate to each scale and resolution of photography. The space photography provides the initial stratification of the landscape into broad macrorelief, primary vegetational, and land-use classes according to the legend. Such broad-scale mapping is illustrated in Color Plate 26(a), showing the Benson-Tombstone-Fort Huachuca and the San Pedro River area on a portion of Apollo frame AS6–1442. This is a gross but highly informative "cut" at macrorelief, vegetational, and land-use mapping for this area. The frame obviously could have been mapped more finely into smaller, essentially "pure" delineations; but, in this instance, the gross features of macrorelief were allowed to control the mapping intensity. Subdivisions were made into large, meaningful areas of pure and complex vegetational and land-use delineations designated at primary and secondary legend level. A treatment such as this can form the basis for aerial subsampling to more precisely define the vegetational and land-use components of the delineations and to develop statistics relevant to these landscape features.

Concentrating on the large, predominantly blue delineation in the Tombstone vicinity (arrow, Color Plate 26(a)), the multistage approach is illustrated. From this initial stratification and interpretation, the area is judged as predominantly undulating to rolling, with some inclusions of hilly lands. (See app. 5–A, symbols 1, 2, and 3.) The undulating to rolling lands generally support vegetation in which whitehorn, creosote bush, and tarbush predominate. (See app. 5–E, symbol 22.2.) Flatlands are of second-order importance in the area. They are suggested by the smooth-textured, pinkish areas and by the dark-blue streaks in the undulating to rolling macrorelief area. The former are yucca grasslands (symbol 43.2); and the latter are narrow bottomland inclusions dominated by dense stands of tobosa grass, with and without mesquite (symbols 43.3 and 43.8). The vegetation of third importance occurs primarily in the rolling lands where stands of mortonia and whitethorn predominate (symbol 22.4). (In the following discussion of vegetation categories 22 and 43, only the units and tenths will be shown as symbols, e.g., 3.8 represents 43.8.)

Although all plant communities known to be in the area do not have unique images on the Apollo

6 photography, more refined mapping can be done from this space photograph, as is illustrated in Color Plate 26(b). Approaching the mountains in the upper-right-hand corner of the illustration, for example, there is an extensive grassland area not separately annotated in Color Plate 26(a). It is characterized by the following species: black and sideoats grama grasses, curly mesquite, and three-awns; with scattered shrubs such as yuccas, mesquite, and Mormon tea.

The outlined square in 26(b) was also covered by one frame of USGS photography taken at an approximate scale of 1:200 000. In an operational multistage survey, these subsamples would be precisely located with respect to random sampling points or transect lines. The small, dashed square in 26(b) corresponds to one frame of still larger scale NASA photography that is essential in the multistage sampling approach.

To perform the first step in refinement of information, the full frame area of 1:200 000 photography was similarly mapped, and each delineation identified (Color Plate 26(c)). The solid-line square in Color Plate 26(c) corresponds to the dashed-line square in Color Plate 26(b). The increased information and mappable detail is easily seen by comparing Color Plates 26(b) and (c). For instance, the highly productive tobosa grass bottomland (designator 3.8) was barely discernible on the space photograph as a thin, dark-blue streak, but it shows in Color Plate 26(c) as a well-defined type. Notice further that on this 1:200 000 photograph, the light-yellow areas within the 3.8 type suggest that there are numerous small inclusions of differing character scattered throughout. A resource area that appeared to be a pure bottomland type in space photography is now confirmed as a complex of two different subjects. It remains for a third stage to discern the true identity of these small yellow inclusions in the darker colored tobosa-grass bottomland.

As more detail becomes necessary to meet land-use and management objectives, the third stage, at scales of 1:20 000 to 1:12 000, is used. At these scales and with adequate sampling intensity, one can obtain useful resource statistics without mapping the entire area of concern at high intensity. This stage is illustrated by 1:1 copies (1:20 000) of parts of the original color transparencies (Color

102

Plates 26(*d*) and (*e*)). From stereo examination of the original 9- by 9-in. transparency, a trained interpreter can identify the most meaningful resource features in the subsample. The effectiveness of this and all previous stages of interpretation is dependent on the adequacy of ground-truth classification into the ecosystems or ecosystem sets responsible for the characteristic images in the photography used at each stage. Thus, the resource analyst is able to work backward through the stages to the space photograph—assuming that all important space photoimages have been subsampled—and define the characteristics of the areas imaged from space with a high degree of statistical and ecological accuracy. Statistics based on either the first or second stages should normally meet the informational needs for broad regional planning and land-use policy formulation. Accuracy requirements for detailed planning generally will require the third stage, and for some applications, acreage determination may require rectification to a planimetric base. For many kinds of resource decisions, a set of statistical data will meet the information needs; but for most resource-management programs, maps are required of the complete management area. These are generally found most useful at scales of 1:31 620 or 1:15 840; and there is a tendency among resource managers to move toward larger scales.

Ground-Truth Documentation

We have demonstrated another use of the legend in connection with Apollo 9 and S065. When doing ground-truth reconnaissance, the legend is particularly suited to the rapid documentation of vegetational and soil surface characteristics. These are the features that provide the basis for interpreting and explaining images on space and other small-scale photography for practical application in vegetational resource management. The system of note taking proved fast enough to use from a small aircraft; and with aerial photographs or topographic map sheets in hand, ground-truth vegetational maps could quickly be made.

In addition to the decimal legend symbol, new components were developed to describe the dominant ground surface features, its cover classes or ground-cover percent; vegetational development stage, level of vegetation removal by animals, and soil surface color. Figure 5–5 shows and explains

three examples of these symbols in terms of the above data classes.

When moving into new areas that are little known, the vegetational community descriptor can be generalized, consistent with one's knowledge; but the other symbols are retained and used as in the example. For instance, highly useful notes can be taken in little-known areas by use of only the primary descriptors or the primary and part of the secondary descriptors. (See apps. 5–C and 5–E and Color Plate 26.)

MULTIDATE INFRARED EKTACHROME PHOTOGRAPHY FOR BROAD-SCALE VEGETATION IDENTIFICATION

Mapping of some of the vegetational groupings itemized in the legend can be facilitated by using Infrared Ektachrome photography. Multidate photography with this film-filter combination emphasizes the differences in seasonal development. Having gained prior knowledge of plant phenologies, the photointerpreter can note the locations of stands dominated by plants with healthy green leaves, which register red on this film-filter combination at each season. Changes among photo dates help to position boundaries and identify images. As seen in Color Plate 27, Apollo 9 Infrared Ektachrome reveals the occurrence of plants with high-infrared reflectance in the mountains during the second week of March 1969. At this time of year and at this geographic location, only the evergreen species show red in the photograph. These species are primarily found in areas identified by legend symbols 65.4, 65.1, 65.2, and 66.1 (app. 5–E). These are the chaparral, oak, and juniper woodlands, and the coniferous forests. The perennial grasses have not begun production of new, green leaves and, therefore, have low-infrared reflectance. Thus this space photograph helps to distinguish the lower elevational limits of these vegetation groups and the upper limits of the desert grassland, because of the color contrast between them at this season. This distinction is not accomplished on the other film types.

On the fringe of the desert, Infrared Ektachrome photography taken at the end of April helps to distinguish the lower elevational boundaries of the desert grasslands (symbol 43.0, app. 5–E), because the cold-season deciduous shrubs of the desert leaf

LEGEND EXAMPLES

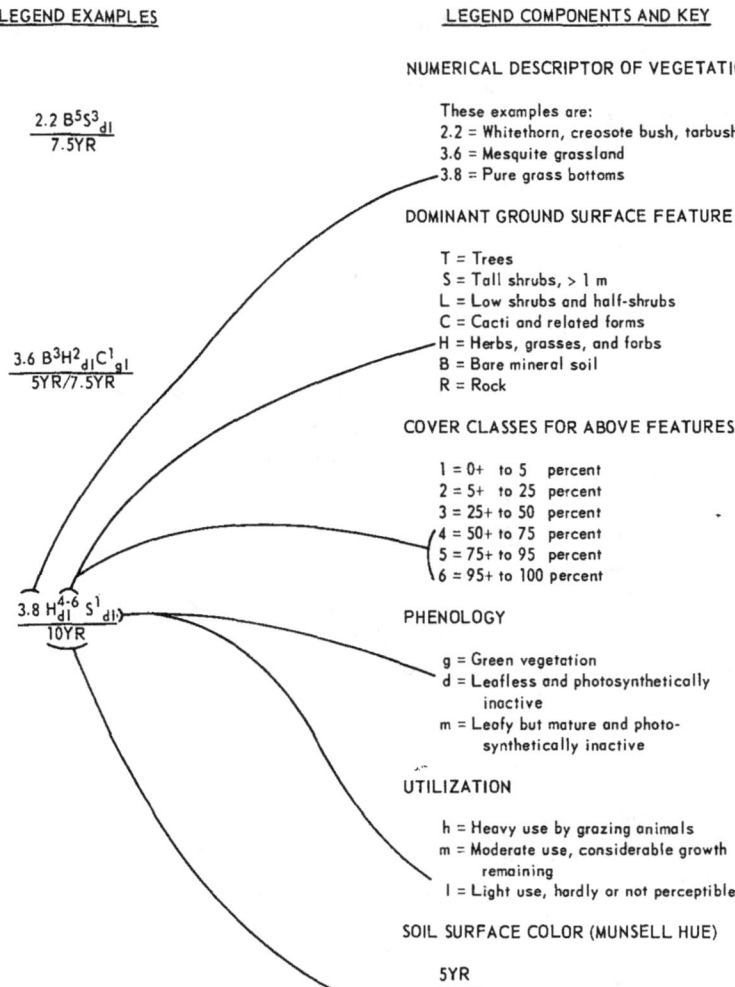

LEGEND COMPONENTS AND KEY

NUMERICAL DESCRIPTOR OF VEGETATION

These examples are:
2.2 = Whitethorn, creosote bush, tarbush
3.6 = Mesquite grassland
3.8 = Pure grass bottoms

DOMINANT GROUND SURFACE FEATURE

T = Trees
S = Tall shrubs, > 1 m
L = Low shrubs and half-shrubs
C = Cacti and related forms
H = Herbs, grasses, and forbs
B = Bare mineral soil
R = Rock

COVER CLASSES FOR ABOVE FEATURES

1 = 0+ to 5   percent
2 = 5+ to 25   percent
3 = 25+ to 50 percent
4 = 50+ to 75 percent
5 = 75+ to 95 percent
6 = 95+ to 100 percent

PHENOLOGY

g = Green vegetation
d = Leafless and photosynthetically inactive
m = Leafy but mature and photosynthetically inactive

UTILIZATION

h = Heavy use by grazing animals
m = Moderate use, considerable growth remaining
l = Light use, hardly or not perceptible

SOIL SURFACE COLOR (MUNSELL HUE)

5YR
7.5YR
10YR
Others as needed

FIGURE 5–5.—Sample legend expressions for rapid ground-truth recording. This legend-notation system can be rapidly and easily used from either the ground or low-flying aircraft once the vegetational characteristics of the region and legend are well learned by the observer. (The symbols 2.2, 3.6, and 3.8 represent 22.2, 43.6, and 43.8, respectively, in app. 5–E.)

out in response to the warming spring temperatures and the winter rains. The warm-season perennial grasses do not begin growing until after the rains of late July and early August. The strong infrared reflectance of the grass during late August shows the extent of the desert grassland when compared with delineations made on earlier photography, and also the extent of good grass understory cover within areas previously identified as desert shrub. Examples of the three generalized groups, that is, evergreens, cold-season deciduous plants, and warm-season deciduous plants, are seen in Color Plate 28. By showing seasonally different views of the same scene, these ground photographs indicate how the three generalized groups of plants can be distinguished on strategically timed photographs. In a similar manner, vegetational types and their aerial photographic images can also be assigned to one of the three generalized groups, according to the phenology of the predominating species. This grouping of vegetations on seasonal development facilitates photorecognition by limiting the alternatives to be considered.

CONCLUSION

The mapping-legend concepts expanded and tested in this experiment for use with space and related aircraft photography appear well suited to the effective analysis, recording, and transmission of relevant information about vegetation and related physiographic features. The flexibility of the legend provides the resource analyst with the means to portray information in a concise form. Data recorded primarily on maps and accompanying descriptive legends enable transfer of varying scales of information for resource planning and management.

High-resolution space photographs, supported by appropriate aircraft imagery, have the demonstrated potential of improving resource inventories. The superiority of Infrared Ektachrome photography for analyzing vegetation resources is clearly evident. The values of multiscale and multiseasonal photography are seen as highly desirable, or even essential, techniques in developing thorough resource inventories of the type studied.

APPENDIX 5–A

MACRORELIEF CLASSES

Symbol	Class name and description
1	*Flatland:* Very gentle slopes, generally under 10 percent; extensive smooth slopes; if interrupted by slopes in excess of 10 percent, these are usually short and represent abrupt changes between different base levels; land may be infrequently dissected by narrow, deep, and steep-sided drainages. The dominant aspect is one of level land.
2	*Smoothly sloping, undulating, and rolling land:* Moderate but smooth slopes in simple systems of slopes and drainages. Slopes are over 10 percent. The general aspect is one of slopes merging smoothly into one another, except in the case of dissected bajadas and pediments. The troughs in the relief pattern tend to return to the same base level (unless rock strata are strongly tilted) rather than for slopes to build upon slopes as in hilly areas. A dissected subclass is often appropriate, indicated by 2.1.
3	*Hilly and strongly dissected land:* Moderate to steep slopes commonly 70 percent, still tend to merge smoothly from pitch to pitch. Ridges tend to be rounded, but the relief pattern is more broken and irregular than class 2. Troughs do not tend to return to a common base level. A moderately complex system of major and minor ridges and swales. Drainage patterns tend to consist of major and minor drainages with the latter extending to higher levels in primary, secondary, and even tertiary patterns; but with the general contour one of smooth relief changes, except in the case of strongly dissected terrain. May include escarpments and cliffs, depending on rock stratification, but these are minor components of the landscape.
4	*Mountainous land:* Moderate to very steep slopes with ridge, slope, and drainage patterns that give a more rugged appearance to the landscape; generally has a higher relief than class 3. Complex major and minor drainage and ridge systems are superimposed one on the other as elevations increase in normally abrupt and steep gradients. Escarpments, rock outcrops, and abrupt changes of slope are more common than in hilly lands. Generally sharp ridgelines and predominantly steep slopes (70 percent) are useful criteria for recognition of mountainous areas.

APPENDIX 5–B

SYMBOLIC MAPPING LEGEND, LANDFORM

Symbol	Ecologically relevant physical features
1.0	Bayous, swamps, tide flats, deltas (vegetated).
2.0	Bottomland, undesignated or unclassified as to type.
2.1	Stringer bottom; narrow, but normally not found in young V-shaped canyons and drainages.
2.2	Valley bottom; wide, including floodplains or "first bottoms."
2.3	Basin, not seasonally ponded.
2.4	Basin, seasonally ponded.
3.0	Alluvial plains, fans, and terraces.
3.1	Bajadas and fans.
3.2	Terraces.
3.21	River.
3.22	Lake.
3.23	Marine.

4.0 Level to rolling uplands, benches, mesas, and plateaus.
5.0 Dunes, sandhills, or beach ridges.
6.0 Slopes; ecologically significant by virtue of a change in vegetation and/or soil with the change in slope.
6.1   Exposed slopes (to prevailing winds and insolation, normally W, SW, S, SE, and sometimes E aspects in Northern Hemisphere; opposite in Southern Hemisphere).
6.2   Protected slopes (from prevailing winds and insolation, normally NW, N, NE, and sometimes E aspect in Northern Hemisphere; opposite in Southern Hemisphere).
6.1 If slopes are ecologically steep, in that they support a different vegetation with the ecotone corresponding to the slope change from moderate to steep, add a designator .01 to the symbol (e.g., 6.11 = exposed, steep slope).
7.0 Patterned ground.
7.1   Biscuit-land complex.
7.2   Ridge-swale complex.
7.3   Pitted-land complex.
8.0 Scabland and/or rockland; vegetated, not barren, >10 percent vegetated.
8.4   On relevant landform feature 4.
8.6   On relevant landform feature 6.
9.0 Ridgetop, convex portion of ridge above tangent with slope, regardless of relative elevation; supports unique vegetation with ecotone more or less at point of tangency; ridge not broad enough to form class 4 feature.
10.0 Canyon, ravine, or arroyo; narrow and deep, young erosionally V-shaped; except arroyos in some soils where they are narrow, vertically sided, and V-shaped.

NOTE.—On Relevant Physical Features 3, 5, and 6, position on slope may be relevant and of ecological significance. When so, as indicated by a change in image characteristics or in vegetation or soils, indicate by .001, .002, and .003 for upper one-third, center one-third, and lower one-third of slope, respectively.

APPENDIX 5–C

SYMBOLIC MAPPING LEGEND, PRIMARY VEGETATION AND LAND USE

Symbol	Physiognomic type or land use
10	Barren land (less than 10 percent vegetated).
20	True deserts (prominent plants scattered; nonvegetated soil surface is dominant landscape feature).
30	Shrub or scrub land (soil surface mostly obscured, shrubs most prominent vegetational feature).
40	Steppes (herbs most prominent vegetational feature).
50	Savannas.
60	Forested and wooded land (arborescent).
70	Alpine or arctic tundra.
80	Vegetation of aquatic environments.
90	Water resources (free water surfaces of mappable size).
100	Agricultural land.
200	Urban and industrial land (including transportational facilities of mappable dimensions).

APPENDIX 5–D

SYMBOLIC MAPPING LEGEND, BARREN LAND

Symbol	Type
1.00	Barren land (<10 percent vegetated).
11.0	Playas.
11.1	Flats, uninterrupted.
11.2	Interspersed with dunes.
11.3	Interspersed with occasional vegetated hummocks.
12.0	Sand dunes.

13.0	Rockland.
13.1	Bedrock outcrops, rimrocks.
13.2	Boulder fields.
13.3	Glacial detritus.
13.4	Lava flow.
13.5	Rock nets, stripes.
13.6	Talus, colluvium.
14.0	Upland barrens (on terraces, plateaus, and undulating lands; not rockland).
14.1	"Badlands," silty /clayey.
14.2	Landslides, fault scarps, erosional escarpments.
14.3	Slicks.
15.0	Shorelines and beaches.
16.0	Made land (raw land resulting from human activity).
16.1	Cuts and fills, nonmining.
16.2	Mining activity.

APPENDIX 5–E

SYMBOLIC MAPPING LEGEND, SECONDARY VEGETATION

Symbol	Vegetational descriptors
21.0	Cactus-microphyll desert.
21.1	Creosote bush with very sparse ground cover.
21.2	Mesquite, creosote bush, burroweed.
21.3	Whitethorn, prickly pear, ocotillo, sparse herbs.
21.4	Saguaro, paloverde, brittlebush, triangle bur-sage.
22.0	Microphyll-thorn scrub desert.
22.1	Whitethorn, mesquite, devoid of herbs.
22.2	Whitethorn, creosote bush, tarbush.
22.3	Mesquite bosques and drainage ways.
22.4	Mortonia, whitethorn.
22.5	Littleleaf sumac, whitethorn, nolina, yucca, white desert zinnia.
22.6	Whitethorn, Wright lippia, ocotillo.
43.0	Steppe.
43.1	Bunch-sodgrass steppe (pure grassland).
43.2	Soaptree yucca grassland.
43.21	Soaptree yucca, sotol grassland.
43.22	Soaptree yucca, sotol, zinnia, coldenia grassland.
43.3	Nolina grassland.
43.31	Mesquite, grama (uplands).
43.32	Mesquite, tobosa grass (bottomlands).
43.4	Mesquite, burroweed grassland.
43.41	Mesquite, burroweed with grasses.
43.42	Mesquite, burroweed, creosote bush without grasses.
43.5	Creosote bush, whitethorn, ocotillo grassland.
43.6	Mesquite grassland.
43.7	Creosote bush grassland.
43.8	Pure grass bottoms.
43.81	Tobosa grassland.
43.82	Sacaton grassland.
54.0	Oak-juniper savanna.
54.1	Oak grassland savanna.
54.2	Juniper grassland savanna.
65.0	Woodland and/or chaparral.
65.1	Oak woodland.
65.11	Oak dominant without tall shrubs.

65.12	Oak with tall shrub layer (chaparral).
65.13	Oak, juniper, grass understory.
65.14	Oak, juniper, chaparral.
65.15	Oak, pinyon pine.
65.2	Juniper woodland.
65.21	Juniper, grassland.
65.22	Juniper, chaparral.
65.23	Juniper, oak, grassland.
65.24	Juniper, oak, chaparral.
65.25	Juniper, oak, pinyon pine, chaparral.
65.3	Pinyon pine woodland.
65.31	Pinyon pine, <5 percent other tree species.
65.32	Pinyon pine, juniper.
65.33	Pinyon pine, juniper, oak, chaparral.
65.4	Chaparral brushland.
66.0	Montane forests.
66.1	Ponderosa pine dominant.
66.2	Douglas fir dominant.
66.3	Engelmann spruce dominant.

NOTE.—This legend is the first iteration of a workable vegetation legend for the Tucson-Willcox-Fort Huachuca triangle of southeastern Arizona.

APPENDIX 5-F

SYMBOLIC MAPPING LEGEND, AGRICULTURAL LAND

Symbol	Type
90.0	Water resources (free water surfaces of mappable size).
91.0	Lakes.
91.1	Natural.
91.2	Artificial or enlarged.
92.0	Watercourses, permanent.
92.1	Rivers and creeks.
92.2	Canals and ditches.
93.0	Bays and estuaries.
94.0	Oceans and seas.

APPENDIX 5-G

SYMBOLIC MAPPING LEGEND, AGRICULTURAL LAND

Symbol	Class
100	Agricultural land.
110	Green and growing crops.
120	Dormant or harvested aftermath.
130	Burned aftermath.
	Uniform subclasses for 110, 120, and 130.
__1	Hay or pasture.
__2	Cereals (excluding corn and sorghums).
__3	Row crops (including corn and sorghums).
140	Orchards, vineyards, cultured forests.
150	Fallow, tilled, or seeded land (not growing).

190	Abandoned land.
191	Revegetating land.
192	Erosional wasteland.

NOTE.—Broad classes of specific crops are indicated under each of the appropriate primary or secondary classes by the tenths and hundredths places, thus: __.00; and the specific crop is indicated by adding one or more digits, progressing toward finer classes (species, variety, etc.) with each progressive digit to the right. Obviously, the farthest digits to the right would tend to require very-large-scale imagery, varietally specific signatures, or ground determination.

APPENDIX 5–H

SYMBOLIC MAPPING LEGEND, URBAN AND INDUSTRIAL LAND

Symbol	Type
200	Urban and industrial land (including transportational facilities of mappable dimensions).
210	Cities and megalopolis.
211	Business districts and shopping centers.
212	Old urban residence.
213	New urban residence.
214	Small-acreage suburban residence.
215	Developing subdivisions and small-acreage suburbia.
220	Towns and villages.
230	Industrial and manufacturing.
290	Transportation developments (surface).
291	Navigable rivers and canals.
292	Major freeways, multiple-lane.
293	Hard-surfaced highways, two- to three-lane.
294	Unsurfaced roads, graded.
295	Unsurfaced roads, ungraded.
296	Railroads.

NOTE.—Ability to use the designators in class 290 obviously depends on the scale and resolution of the imaging system.

APPENDIX 5–I

SPECIES LIST

Common name	Scientific name
Brittlebush	*Encelia farinosa* Gray
Burroweed	*Haploppapus tenuisectus* (Greene) Blake
Bur-sage, triangle	*Franseria deltoideae* Torr.
Coldenia	*Coldenia canescens* D.C.
Creosote bush	*Larrea tridentata* (D.C.)
Fir, Douglas	*Pseudotsuga menziesii* (Mirbel) Franco.
Grama	*Bouteloua* sp.
Juniper	*Juniperus* sp.
Lippia, Wright (honeysage)	*Aloysia wrightii* (Gray) Heller
Mesquite	*Prosopis juliflora* var. *velutina* (Wool.) Sarg.
Mormon tea (jointfir)	*Ephedra trifurca* Torr.
Mortonia	*Mortonia scabrella* Gray
Nolina (beargrass, sacahuista)	*Nolina microcarpa* Wats.
Oak	*Quercus* sp.
Ocotillo	*Fouquieria splendens* Engelm.

Paloverde	*Cercidium* sp.
Pine, pinyon	*Pinus edules* Engelm.
Pine, ponderosa	*Pinus ponderosa* Lawson.
Prickly pear	*Opuntia Engelmannii* Salm-Dyck.
Sacaton	*Sporobolus Wrightii* Munro.
Saguaro	*Carnegiea gigantea* (Engelm.) Britt. & Rose
Sotol	*Dasylirion wheeleri* Wats.
Spruce, Engelmann	*Picea Engelmannii* Parry.
Sumac, littleleaf	*Rhus microphylla* Engelm.
Tarbush	*Flourensia cernua* D.C.
Three-awn	*Aristida* sp.
Tobosa	*Hilaria mutica* (Buckl.) Benth.
Whitethorn	*Acacia constricta* Benth.
Yucca	*Yucca arizonica* McKelvey
Yucca, soaptree	*Yucca elata* Engelm.
Zinnia, white desert	*Zinnia pumila* Gray

6

Evaluation of Wildland Resources of the NASA Bucks Lake Test Site

Andrew S. Benson

The 100 000-acre NASA Bucks Lake test site on the west side of the Sierra Nevada Mountains of California has been studied extensively by the Forestry Remote Sensing Laboratory of the University of California for 5 years. The site contains a wide variety of wildland resources including timber, which is the primary factor in the economy of the area; brush, which provides the habitat for wildlife but in some instances might profitably be replaced by timber; forage suitable for grazing by domestic livestock; water, which is rapidly becoming the most important resource of the area; and both present and potential recreation areas. Elevations within the test site vary from 3600 to 7000 ft, thus providing a number of climatic regimes that are reflected in differing soil and vegetation types (Draeger, 1967).

High-altitude photography of an area such as this (i.e., from an altitude of 60 000 to 75 000 ft) can aid a wildland manager in (1) mapping types of ground cover (vegetation), (2) monitoring changes in the ground cover with sequential coverage, and (3) selecting primary sample units for more detailed study in multistage sampling. (See ch. 8.)

Conventional low-altitude photography (i.e., from an altitude of 10 000 to 15 000 ft) also can be used for these tasks, but fewer high-altitude photographs are needed to cover a given area. The entire Bucks Lake test site, for example, was covered with one flightline of four high-altitude photographs (60 percent overlap); similar coverage with 9- by 9-in. low-altitude photographs (1/20 000) would require at least three flightlines of 10 photographs each.

GROUND-COVER MAPPING

For ground-cover mapping, three types of high-altitude photography were compared: Infrared Ektachrome, Infrared Aerographic with a Wratten 89B filter (IR-89B), and Panatomic-X with a Wratten 25A filter (Pan-25A). The original 35-mm photography was enlarged to produce 5- by 7-in. prints having a scale of 1/150 000. The photography obtained with each film-filter combination was viewed stereoscopically, and the cover-type boundaries were drawn. The delineations were transferred to an Infrared Ektachrome mosaic (Color Plate 29) and single black-and-white photographs (figs. 6–1 and 6–2) for comparison.

Ground cover was typed according to six categories (160 acres, minimum area): (1) brush or dry-site hardwood, (2) medium- to high-density conifer, (3) low-density conifer, (4) rock or bare soil, (5) meadow or riparian hardwood, and (6) water. The advantages of high-altitude photography for delineating these categories is quite apparent. Because so few photographs are required to cover the entire area, the photographic tones and textures used in identifying any given type of ground cover tend to be consistent throughout most of the area, and the number of photographs to be handled is small. In an area like Bucks Lake, the mapping of ground-cover types on photographs of conventional scale (e.g., 1/20 000) would introduce much greater variability in photographic tones and textures of any given type of ground cover, and would require the shuffling of 30 or more photographs. This is not only time consuming but interrupts the concentration of the interpreter.

Of the three types of photography used, Infrared Ektachrome proved to be the most satisfactory for cover typing (Color Plate 29). In most instances each category can be easily distinguished from the others. However, some confusion does occur between dry-site hardwoods and riparian hardwoods, and between low-density conifers and brush. In

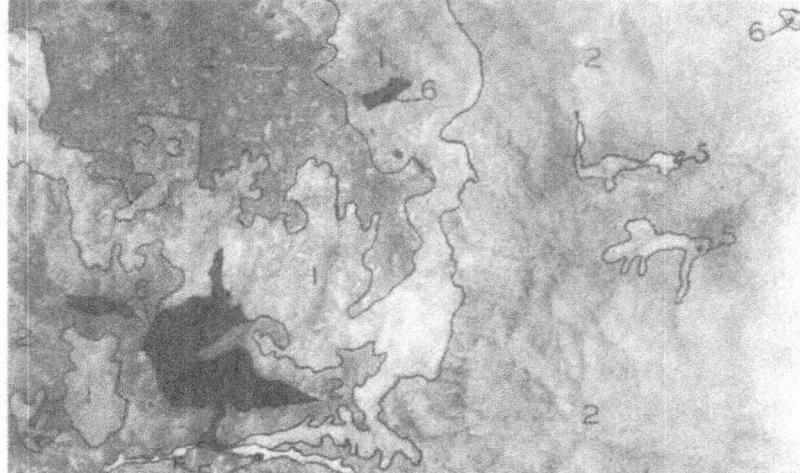

FIGURE 6-1.—IR-89B photography taken on July 15, 1969, from an altitude of 70 000 ft. This type of photography proved to be of little use for ground-cover typing. Except for bodies of water and riparian hardwoods, most of the above delineations and identifications are of questionable accuracy, as indicated by comparing this figure with Color Plate 29. (See legend on Color Plate 29.)

the first case, the confusion between dry-site hardwoods and riparian hardwoods can be resolved when the photographs are viewed stereoscopically, because the lowland topography occupied by riparian hardwoods is quite different from the upland topography of dry-site hardwoods. In the second case, because young conifers, unlike mature conifers, display high-infrared reflectivity they can be confused with brush, particularly when the young conifers are emerging from beneath a brush overstory. The young conifers are distinguishable on large-scale photography only where the tips of the emerging trees can be seen. This condition oc-

curs at area A (Color Plate 29) and resulted in the only major mapping error of ground-cover boundaries on the high-altitude photography. Examination of this area on large-scale photography indicated that area A is low-density conifer and not brush or dry-site hardwood. Despite this confusion, it should be noted that the resolution of the Infrared Ektachrome high-altitude photography is good enough to allow mapping of features less than 160 acres in size.

Judging from this study, IR-89B photography (fig. 6-1) is of little use for cover typing. Bodies of water, watercourses, wet meadows, and riparian

FIGURE 6-2.—Pan-25A photography taken on July 15, 1969, from an altitude of 70 000 ft. Ground-cover typing can be done fairly consistently on this photography although not as readily as on the Infrared Ektachrome photography of Color Plate 29. Some difficulty occurs, for example, when one attempts to distinguish between categories 2 and 3, and also between 3 and 4. (See legend on Color Plate 29.)

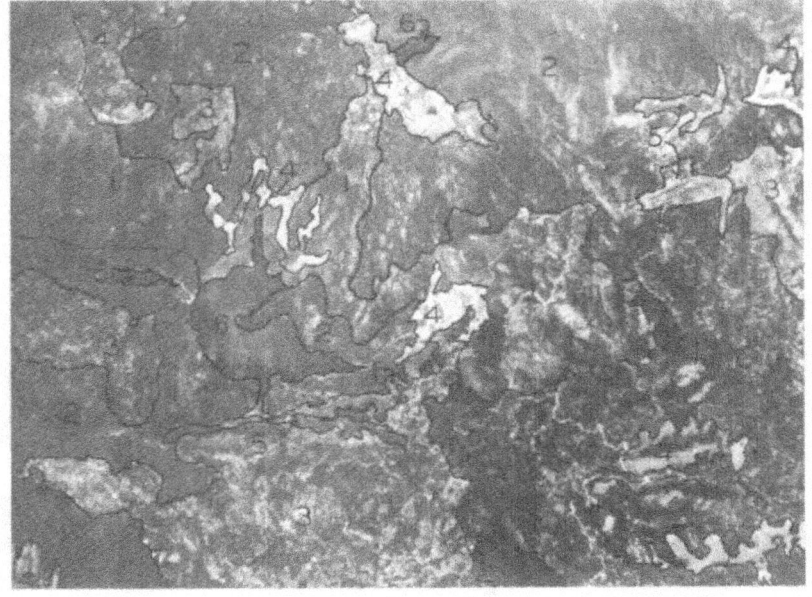

hardwoods are the only cover types that can be distinguished. Brush fields cannot always be separated from the rock-bare-soil and/or low-density-conifer categories. The two density classes of conifers are indistinguishable. Finally, the road network is not visible; thus the land manager has no indication of the location of his resources with respect to available transportation.

Cover typing can be done satisfactorily with Pan-25A photography (fig. 6–2), but not as easily as on Infrared Ektachrome. The gray tones of meadow and riparian vegetation are quite similar to those of brush and dry-site hardwood vegetation. When the photographs are viewed stereoscopically, however, the topography can be used to differentiate the two categories. Whereas bare soil is easily identified, there can be some confusion between areas of bare rock and areas of low-density conifer.

MONITORING CHANGES IN COVER PATTERNS

Sequential high-altitude photography gives the land manager a permanent record of changes in ground cover. The most evident example of this has been seen at *B* in Color Plate 29. June 1966 photographs show no bare soil at *B*, but the outline of a brushland-conversion site is clearly seen at *B* in Color Plate 29. If this area is photographed again in about 10 yr, a comparison with the present photography may indicate the results of this manipulation of the vegetation by man.

Sequential high-altitude photography is ideal for monitoring changes in snowpacks. The frequency of the photography required varies. Comparison of the snowpack on the May 21, 1969, high-altitude photography (fig. 6–3) with the high-altitude photography obtained on July 15, 1969 (fig. 6–2), indicates that a 2-month interval is too long to be of maximum use for snow surveys. Biweekly photography from December 1 to June 1 would be desirable. During the periods of maximum runoff, weekly photography might be more desirable (Sattinger and Polcyn, 1966). The sequential photography, when coordinated with pertinent ground data, would provide a permanent record of snow accumulation and melt patterns for a given area. This, when combined with data from previous years, would enable a land manager to

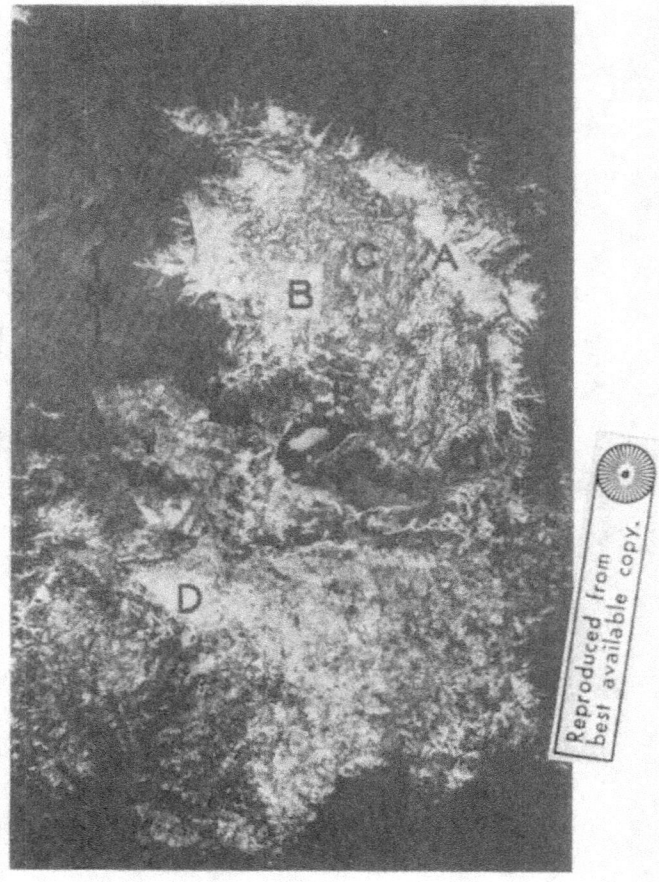

Legend

A Steep (50 percent) north-facing slope devoid of woody vegetation High snow accumulation and retention.

B Gentle south-facing slope with low-density-conifer cover. High snow accumulation and low retention.

C Moderately dense conifer cover on gently rolling topography. Low snow accumulation and retention.

D Moderately steep north-facing granitic outcrop. High snow accumulation and low retention.

FIGURE 6–3.—Pan-58 photography taken on May 21, 1969, from an altitude of 70 000 feet. Relative accumulation and retention of snow with respect to aspect, slope, and ground cover can be discerned on this photograph.

predict the timing and relative quantity of water yield from the area (Draeger, 1968).

Snow-accumulation patterns, when compared with ground-cover patterns, slopes, and aspects, give a land manager an indication of how the manipulation of vegetation cover, such as brush-

114

field reclamation and heavy-timber harvesting, affect the distribution and accumulation of snow. Figure 6–3 provides four examples of these relationships, as follows:

(A) This steep (50 percent) north-facing slope is devoid of woody vegetation. Snow accumulation is relatively high because it faces to the north, where the Sun's rays strike it only obliquely. The snow is retained here until late in the spring. Consequently, the growing season each year is so short that only herbaceous plants can grow here. Such an area is known as a "snow glade."

(B) Snow accumulation is high in this area also, but snow retention is less persistent because the slope faces south. This area has a very low density of conifers because timber harvesting was undertaken ostensibly to convert the land from timber production to future vacation-home sites.

(C) This area has the same slope and aspect as area B, but the vegetation cover is more dense. Snow accumulation and retention are low.

(D) Although this granitic outcrop is predominantly north facing, its snow accumulation and retention are less than that of area A because the elevation is lower and the slope is less steep. The near absence of vegetation in this area is attributable to edaphic rather than climatic factors.

High-altitude photography and even space photography (see ch. 3) can give the areal coverage and the relative accumulation of the snowpack, but ground measurement is still required to determine snow volumes. The determination of the water content of snow using active remote sensors is still in the experimental stage (Meier, 1968); and while snow depths can be determined photogrammetrically, the optimum photographic scale for this is 1/6000 (Cooper, 1965).

APPLICATIONS FOR MULTISTAGE SAMPLING

In chapter 8, examples are given of how either space photography or high-altitude aerial photography can be used in selecting primary sample units as the first step in multistage sampling. Since more detail can be seen on the high-altitude photography, the sampling errors can be expected to be lower than if space photography were used. More study is needed to determine whether, with high-resolution high-altitude photography, this particular design need only consist of two or three

stages when applied to inventory problems of the types discussed in this chapter.

Multistage sampling can be used effectively for inventorying almost any wildland resource including timber, forage, water, and snow. Depending upon the resource being inventoried and its distribution (i.e., clustered or homogeneous), a mosaic such as the one appearing in Color Plate 29 can be partitioned into squares of arbitrary size, perhaps 1 or 2 mi on a side. These squares then become the population from which the primary sampling units are to be selected. The resource is delineated on the entire mosaic, and the proportion of each square occupied by it is estimated by the interpreter. The primary samples are then selected with probability proportional to the prediction with the use of random-number tables.

CONCLUSIONS

High-altitude photography can be extremely useful in evaluating wildland resources. Because so few photographs are needed to cover a given area as compared to conventional low-altitude photographs, evaluation is simplified. Infrared Ektachrome was judged best for delineating ground-cover boundaries, although Pan-25A was satisfactory for delineating several cover types. Except for delineating bodies of water, watercourses, and riparian hardwoods, IR–89B proved unsatisfactory for overall ground-cover typing. Sequential photography, if taken over a number of years, will yield valuable information concerning the consequences of vegetation manipulation and the yield of snow. More study is needed, however, to determine how high-altitude photography can best be used in conjunction with conventional photography and/or space photography to help inventory wildland resources.

SELECTED LITERATURE

COOPER, C. F. 1965. Snow Cover Measurement. Photogram. Eng. 31(4): 611–619.
DRAEGER, W. C. 1967. The Interpretability of High Altitude Multispectral Imagery for the Evaluation of Wildland Resources. Annual Progress Rept. Earth Resources Survey Program, NASA.
DRAEGER, W. C. 1968. The Interpretability of High Altitude Multispectral Imagery for the Evaluation of Wildland Resources. Annual Progress Rept. Earth Resources Survey Program, NASA.
LANGLEY, P. G.; R. C. ALDRICH; AND R. C. HELLER.

1969. Multi-Stage Sampling of Forest Resources by Using Space Photography—An Apollo 9 Case Study. Vol. 2: Agriculture, Forestry, and Sensor Studies. Proc. 2d Annual Earth Resources Aircraft Program Status Rev., pp. 19–1 to 19–21. NASA, MSC, Houston, Tex.

MEIER, M. F. 1968. Evaluation of South Cascade Glacier Test Site. Vol. 3: Hydrology, Oceanography and Sensor Studies. Earth Resources Aircraft Program Status Rev., NASA, MSC, Houston, Tex.

SATTINGER, I. J.; AND F. C. POLCYN. 1966. Peaceful Uses of Earth Observation Spacecraft. Vol. 2: Surveys of Applications and Benefits. IROS, Willow Run Laboratories, Inst. Sci. Tech., Univ. of Michigan, Ann Arbor.

Color Plate 25. This illustration shows most of frame AS6–1442. Each delineation represents broad-scale, vegetation-macrorelief units. The symbols of the numerator and denominator represent the primary vegetation and macrorelief classes, respectively. Details are given in the text. The mapped area includes all of the study triangle. Willcox Playa is clearly evident in the upper right-hand corner of the photograph as a white irregular basin (11.1). Fort Huachuca, in the bottom center of the photo, is barely discernible. Tucson, in the upper left-hand corner, is indicated by the descriptor 200.

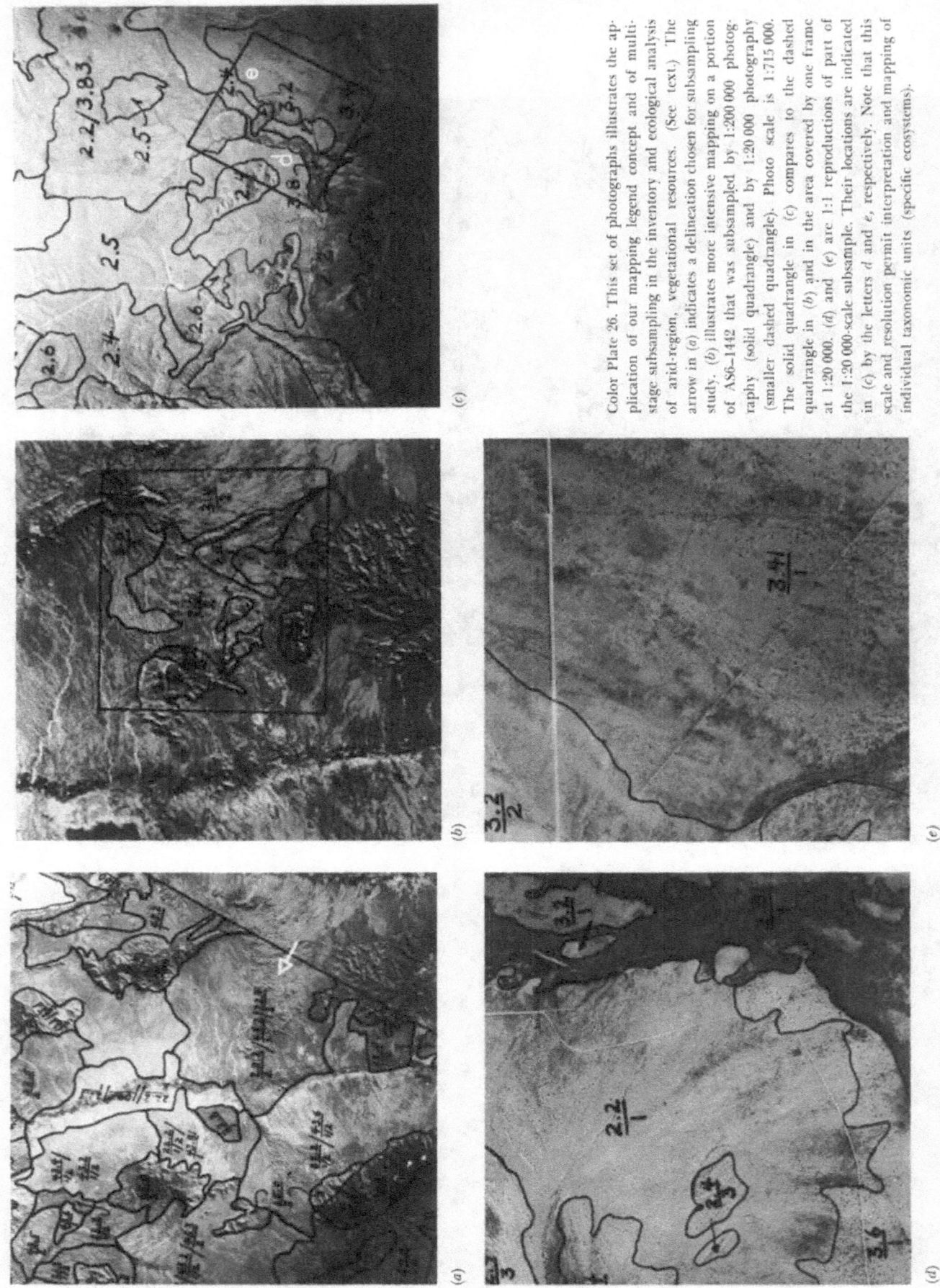

Color Plate 26. This set of photographs illustrates the application of our mapping legend concept and of multistage subsampling in the inventory and ecological analysis of arid-region, vegetational resources. (See text.) The arrow in (a) indicates a delineation chosen for subsampling study. (b) illustrates more intensive mapping on a portion of A86–1442 that was subsampled by 1:200 000 photography (solid quadrangle) and by 1:20 000 photography (smaller dashed quadrangle). Photo scale is 1:715 000. The solid quadrangle in (c) compares to the dashed quadrangle in (b) and in the area covered by one frame at 1:20 000. (d) and (e) are 1:1 reproductions of part of the 1:20 000-scale subsample. Their locations are indicated in (c) by the letters d and e, respectively. Note that this scale and resolution permit interpretation and mapping of individual taxonomic units (specific ecosystems).

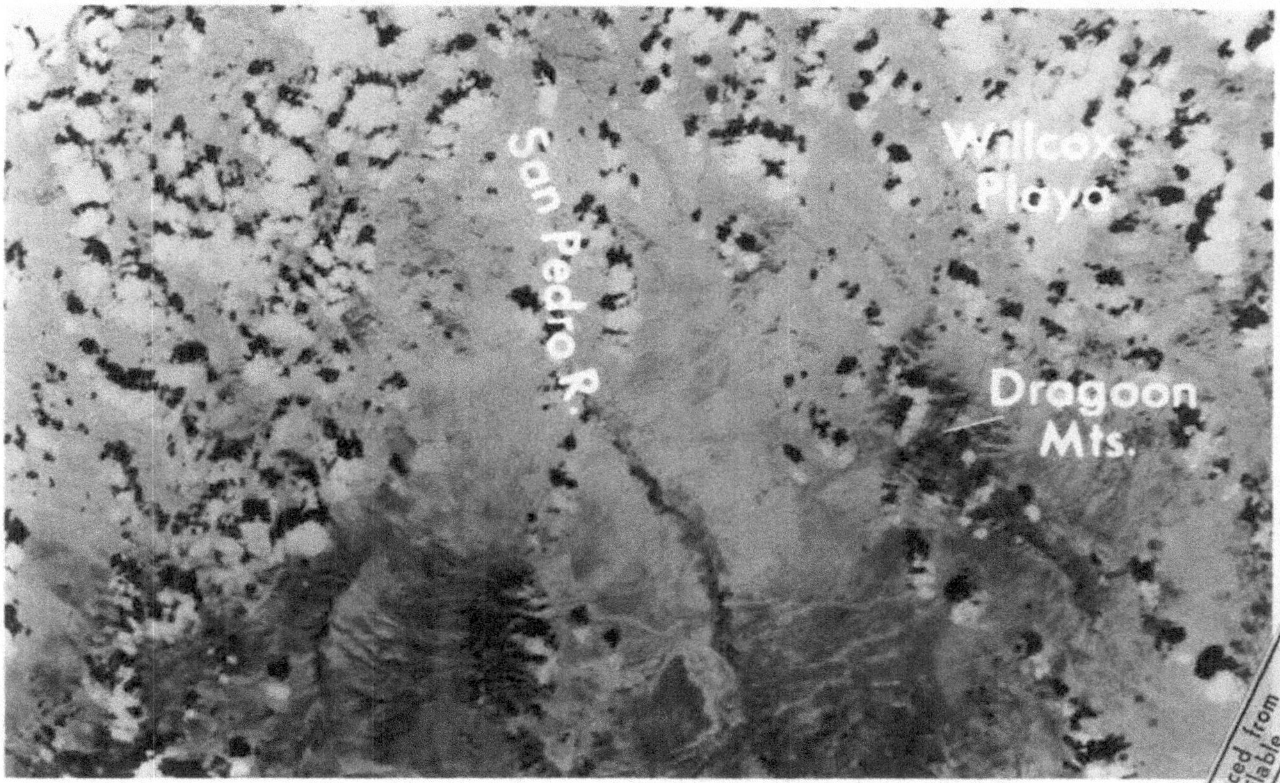
San Pedro R.
Willcox
Playa
Dragoon
Mts.

Color Plate 27. Seasonal variation in the color infrared signature of native vegetation communities aids in their accurate delineation and recognition. This photo is an enlargement from an Apollo 9 frame taken during the second week of March 1969. At that time, only the evergreen species had green foliage and therefore registered red in color on this film-filter combination. Note that these species predominate in the higher mountain ranges where an increase in the red intensity is evident. Careful examination of the original photo reveals the boundaries of these woody, evergreen vegetation types rather distinctly and sets them apart from the dormant vegetation of the desert.

April 30, 1969

May 22, 1969

August 30, 1969

September 30, 1969

Color Plate 28. These Infrared Ektachrome ground photos depict the same scene on different dates. The plants include evergreens (yuccas, Y), cold-season deciduous desert shrubs (mesquite, P), and warm-season deciduous species (perennial grasses, G; whitethorn, A). In late April, only the yuccas and chaparral had predominant green foliage and appeared red. In late May, mesquite had leafed out and retained its green color throughout the rest of the photo dates. In late August, the perennial grasses and whitethorn appeared red, but the grasses lost their green foliage faster than the whitethorn. Thus, in late September, the difference between these two warm-season deciduous types was evident.

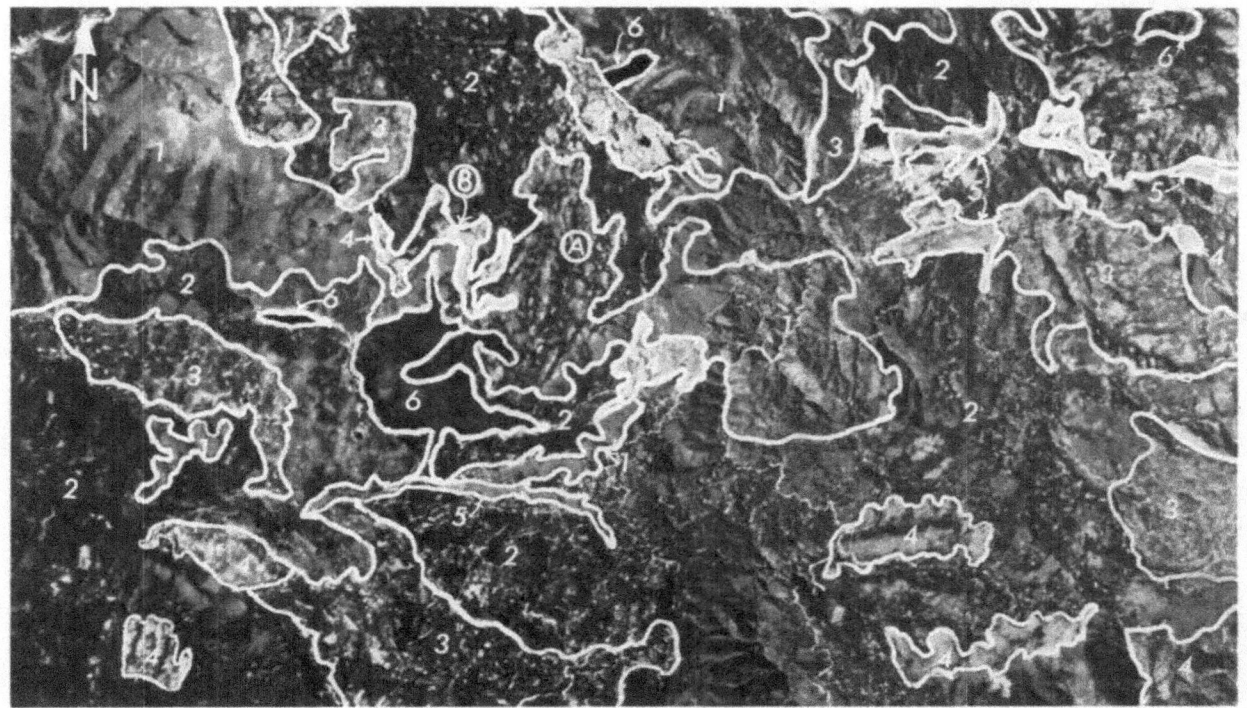

Legend

1	Brush or dry-site hardwood	4	Rock or bare soil
2	Medium- to high-density conifer	5	Meadow or riparian hardwood
3	Low-density conifer	6	Water

Color Plate 29. This mosaic of the NASA Bucks Lake Test Site was made from three 35-mm Infrared Ektachrome photographs that were taken at an altitude of 70 000 ft on August 5, 1969. Of the three types of photography tested, Infrared Ektachrome proved to be the most satisfactory for cover typing. One major mapping error occurred at area *A*. Examination of this area on large-scale photography indicated that area *A* is low-density conifer and not brush or dry-site hardwood. See text for further discussion.

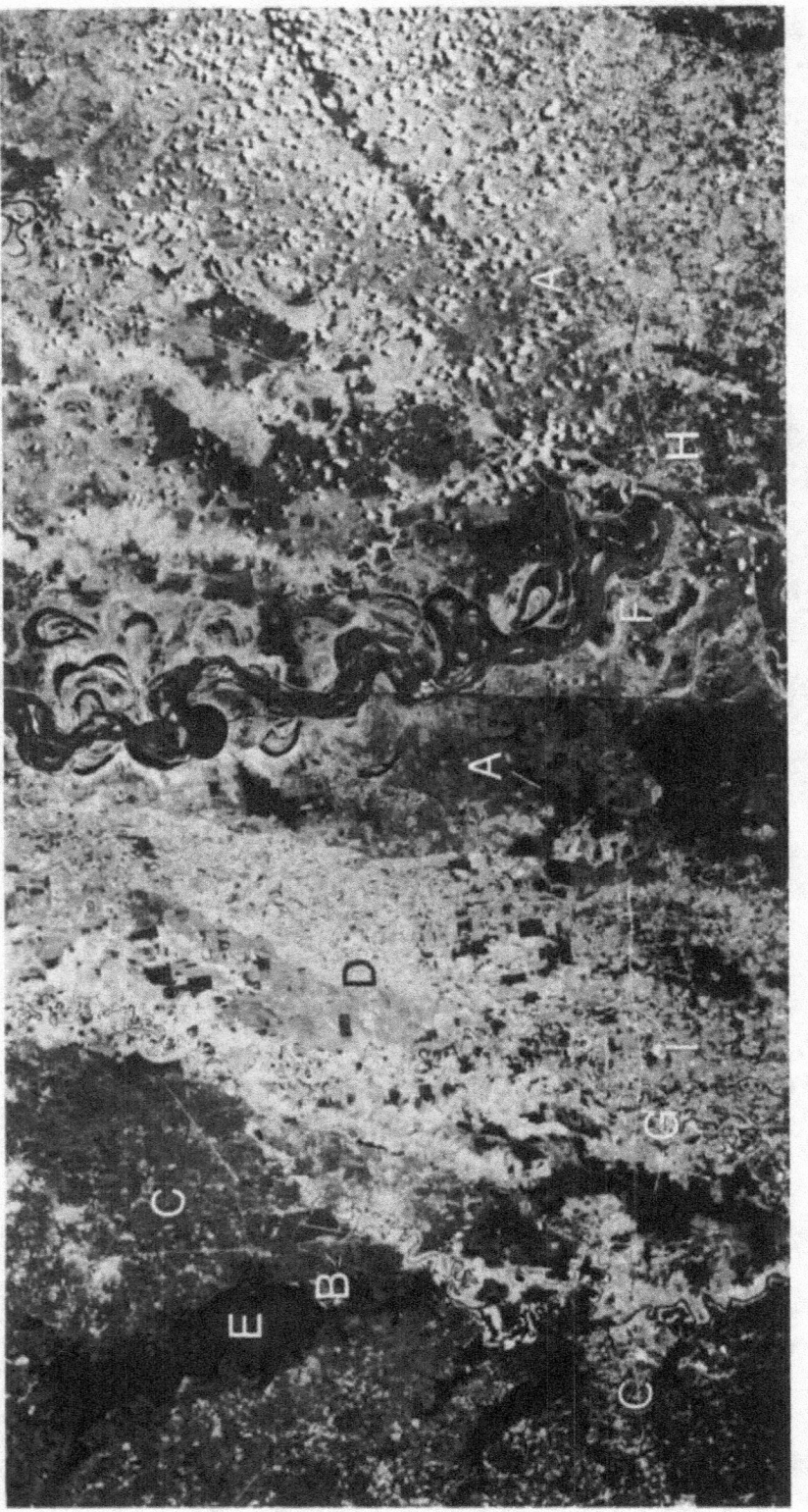

Legend

A	Deciduous forest	F	Rivers
B	Pine forest	G	Canals
C	Mixed deciduous and pine forest	H	Urban and industrial area
D	Cultivated land	I	Major roads
E	Open water		

Color Plate 30. The photographs that comprise this mosaic were taken in March 1969 on Infrared Ektachrome film and have been enlarged two diameters from a copy of the original space photo.

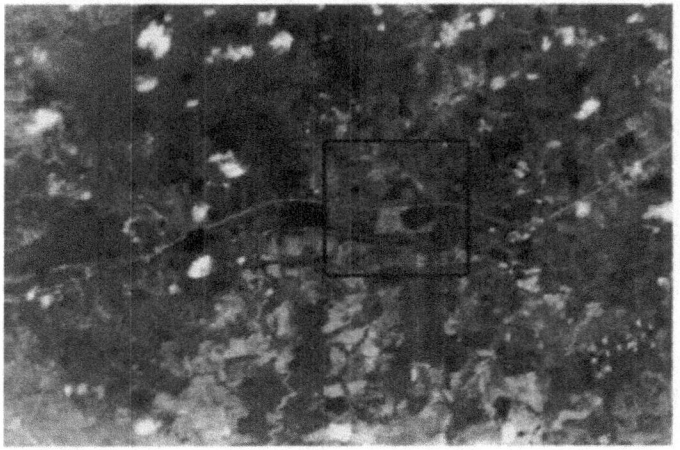

Pan-25A (space)

Color Plate 31. Deciduous forests, pasture, cultivated land, roads, water courses, and urban features can be identified on the enlargements of space photographs (*a*) and (*b*). The Infrared Ektachrome vertical photo (*c*) was taken at an altitude of 20 000 ft by a K–17 camera (focal length = 12 in.). The area it covers is outlined on space photographs (*a*) and (*b*). It shows deciduous forests intermixed with pasture and cultivated land. Note the areas where deciduous forests stand in water at the center of the photo (darker blue areas). The oblique aerial photo (*d*) shows a portion of the area outlined in (*a*) and (*b*) and seen in (*c*). Comparison of detail in the oblique with the aerial and space photos will confirm the identity of the land-use categories.

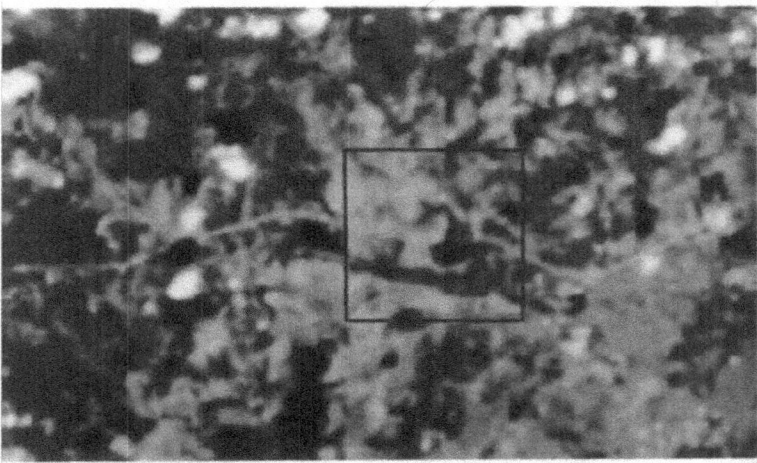

Infrared Ektachrome (space)

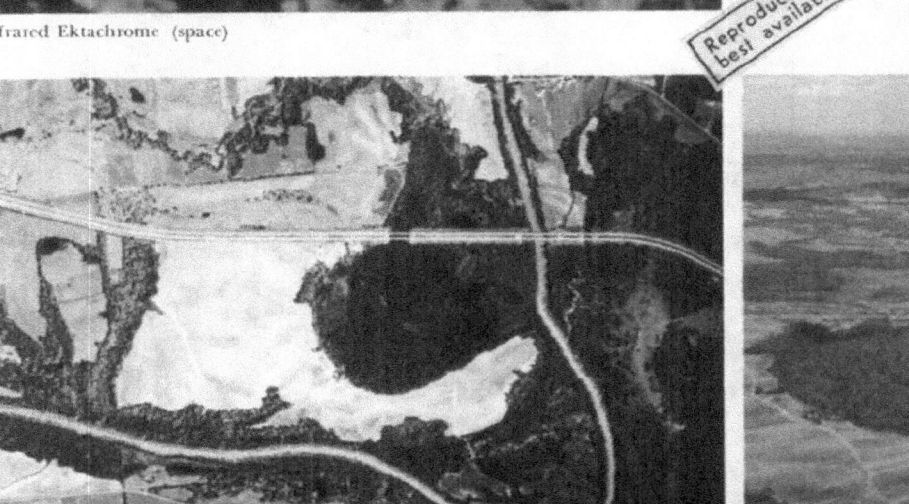

Infrared Ektachrome (high-altitude vertical)

(*d*) Oblique aerial photograph

March 9, 1969 (Apollo 9 enlargement); Infrared Ektachrome

Color Plate 32. On an enlargement of a portion of AS9–26A–3740 (*reproduced above*) a feature identification test was administered. The area is just west of Monroe, La. The two aerial oblique photographs (*at the left*) cover the portion of the Apollo 9 photo in which features 2 through 6 are annotated. These annotations indicate locations of examples of the land-use categories that were chosen for the test: (1) deciduous forest, (2) pine forest, (3) mixed deciduous and pine forest, (4) cultivated land (bare), (5) cultivated land (vegetated), (6) cultivated land (fallow), (7) open water, and (8) major road (freeway). The aerial oblique photograph at the top was taken 2 weeks after the Apollo 9 photo. Vegetation appears in the oblique photo in essentially the same condition as when photographed by the astronauts. The oblique photograph at the bottom was taken on November 11, 1969, when the deciduous hardwood trees were undergoing the fall color change. The value of sequential photography for mapping vegetation is discussed in the text (ch. 7).

March 24, 1969; Infrared Ektachrome

November 11, 1969; Infrared Ektachrome

7

A Land-Use Classification System From Apollo 9 Photographs for the Mississippi-Louisiana Area

ROBIN I. WELCH, LAWRENCE R. PETTINGER, AND EDWIN H. ROBERTS

A mix of aerial and space photography is far more useful for Earth resource surveys than either type alone if three requirements are met: (1) the photography is of the "multiband" type, (2) the photorecognition characteristics of each kind of Earth resource are set forth in some clear fashion such as a photointerpretation key, and (3) ground data are collected at representative locations throughout the area.

The Apollo 9 astronauts, using the S065 camera system, obtained excellent multiband photography of the Mississippi-Louisiana area at a time when the area was unusually cloud free. Shortly thereafter, personnel of the Forestry Remote Sensing Laboratory visited the area, took aerial photos of selected portions of it, and also acquired ground data at many representative spots. This information was used to enhance the value of the space photography and also to determine the accuracy with which land-use classifications might be made in this area from such photography.

With modern multiband photographic systems it is possible to acquire so much aerial and space photography in a very short time that one can be overwhelmed by it. Therefore, it is imperative for the wildland resource analyst to determine beforehand just what kinds of photos he needs and request only those he is prepared to interpret.

On aerial photography flown to specifications, it is possible to (1) identify timber and agricultural crop species, (2) detect plant diseases, (3) estimate timber volumes and crop yields, (4) evaluate range resource conditions, (5) inventory wildlife, (6) delineate vegetation and soil boundaries, (7) evaluate wildlife habitat, and (8) perform many other tasks. On space photography none of the above can be accomplished as well as on aerial photography, but information such as the following frequently can be obtained: (1) regional terrain and vegetation patterns, (2) regional weather patterns, (3) condition of watercourses and large bodies of water, (4) regional soil boundaries obscured by erosion or vegetation cover, (5) regional snow cover, and (6) similar data that would otherwise require instantaneous photography from an enormous number of aircraft and subsequent mosaicing of thousands of photos.

Ground observations are required to quantify and verify information derived from photointerpretation activities. An interpreter can rapidly delineate soil, terrain, and vegetation patterns on space photography, however; thus he can direct ground crews to promising areas and eliminate other areas from consideration.

Generalizations such as those in the preceding paragraphs can be useful. However, they become much more meaningful when (a) a specific geographic area is studied in detail, as in this chapter, (b) meaningful categories of land use are applied to that area, (c) the photorecognition characteristics for each category are set forth in some systematic fashion, (d) a quantitative determination is made of the accuracy with which each category can be recognized on photography flown to proper specifications, and (e) for each category an analysis is made of the types of commission and omission errors that are made by the photointerpreters. All these elements are contained in the present chapter as applied to land-use classification in the Mississippi-Louisiana area.

CLASSIFICATION SYSTEMS FOR FOREST LAND

A classification system devised for the S065 photography of the area surrounding Vicksburg, Miss.,

118

recognizes the following land and vegetation features:

(1) Deciduous forest
(2) Pine forest
(3) Mixed deciduous and pine forest
(4) Cultivated land
    (*a*) Bare ground
    (*b*) Vegetation-covered ground
    (*c*) Fallow ground
(5) Open bodies of water (lakes, reservoirs, etc.)
(6) Rivers and canals
(7) Urban and industrial areas
(8) Major roads

Although these classes may be further subdivided by inspection of aerial photographs of the same areas, it does not seem feasible to further subdivide identifications made from space imagery because (1) errors will undoubtedly occur that will reduce the overall accuracy of the interpretation and (2) the increased time taken by an interpreter to make these delineations will not be justified. Most projects will have access to concurrent aircraft photographs of portions of the total area; and by using suitable sampling methods, interpreters can assemble the necessary data for subdividing classes and quantifying land-use types delineated on space imagery.

Many of the criteria used for recognizing features on aerial photographs can also be applied to space photographs; however, the interpreter of space photos sees land units the size of city lots and forests rather than individual trees and shrubs. Hence, he must change his scale of thinking. The color and texture of images may be altered severely by the view from space. One may continue to rely on the classical image characteristics of color or tone, shape, size, texture, and location of features used in identifying images taken from aircraft for the new field of space-image interpretation, but it may not be possible to correlate several of these criteria because of limitations imposed by small image size and limited spatial resolution. If aerial photography of portions of the area covered by a space photo is available, the image-interpretation task may be considerably easier.

The photos considered in this section provide examples of recognition features of forest areas and other scene components occurring on S065 photography taken in Mississippi and Louisiana. The photorecognition characteristics for the land and vegetation categories of this area are summarized in table 7–1.

Deciduous Forests

The seasonal state of deciduous forests strongly influences the appearance of such areas. At the time of S065, the deciduous trees were dormant without foliage. Under these conditions, differentiating deciduous forest areas from coniferous forests and pasture areas was relatively easy on aerial photography and on space photography where forest areas were large enough to be resolvable. On Infrared Ektachrome photos, deciduous forests are dark blue to black and appear as a nearly uniform color. (See Color Plate 30 and compare with Color Plate 31.)

Pine Forests

At the seasonal state represented in S065, pine forests contrast with deciduous forests because of the absence of foliage on deciduous trees. Pine forests can be seen on the aircraft photos, and where a block of pine forest is large enough, it is also visible on space photos.

The color of pine forests on Infrared Ektachrome aerial photos varies from pink to dark red depending upon age and stand density. On space photos, pine forests appear only slightly reddish in color. Where a forest contains both pine and deciduous trees, it is frequently not possible to detect pine trees because they occur in small groups and their reddish signature is not visible. These areas will be classified as deciduous trees on S065 space photos. This situation would probably hold true in summer when all trees are in foliage. Under these circumstances supporting aerial photography is very helpful. (See Color Plates 30 and 31 and fig. 7–1.)

Mixed Deciduous and Pine Forests

Many of the forest areas to be delineated on the S065 photographs of the Mississippi-Louisiana area are classified as supporting both deciduous and pine forests. Where aircraft photos are available, many of the areas can be accurately classified according to timber types. These observations can be extended to adjacent areas on space photos using the recognition features presented. For mixed forests, location becomes important in identification. The rolling terrain east and west of the river course and flood plain generally supports mixed

TABLE 7–1.—*Recognition Characteristics for Land and Vegetation Categories in the Mississippi-Louisiana Study Area on Apollo 9 Space Photography*

Land-use or vegetation-type category	Characteristic, by film-filter combination [a]		
	Infrared Ektachrome	Pan-25A	IR–89B
Deciduous forests—in large blocks along rivers and on flood plains.	Dark blue to black (variable tones).	Dark gray to black	Medium gray
Pine forests—in small blocks on uplands.	Dark reddish patches (variable tones).	Dark gray to black	Medium gray
Mixed deciduous and pine forests—in uplands.	Dark blue to black and occasionally reddish (variable tones).	Dark gray to black	Medium gray
Cultivated land—in flood plain and uplands.	Bare: light tan to white Vegetated: light red to pink Fallow: light blue-gray (uniform tones in most fields).	Light gray to white	Light gray
Open bodies of water (rivers and canals).	Mostly uniform blue to black	Light gray to black	Medium gray to black
Urban and industrial areas—streets, buildings, airports, and highways.	Light-toned to black	Light gray to dark gray	Light gray to white
Major roads—linear features with intersections.	Tan to white	Light gray to white	Light gray to white

[a] Pan-25A is Panatomic-X film with a Wratten 25A filter; IR–89B is infrared film with a Wratten 89B filter.

forests, while the flood plain contains mostly pure deciduous forests. It is difficult to use color and texture alone for delineation of deciduous versus pine forests on space photos unless large pure stands of either type are present. Large plantations of pine forest usually are identifiable at this time of year by their reddish color on Infrared Ektachrome photography, together with their texture and topographic location.

Cultivated Land

Cultivated land (i.e., bare fields, irrigated pasture land, fallow fields, and fields containing agricultural crops) is largely confined to the flood plain except for a mixture of forest and small areas of cultivated land in the uplands. On space photos cultivated fields often appear to be relatively variable in tone one to the other but uniform in shape.

Little difficulty is encountered in separating cultivated land and forest land on space imagery as seen in Color Plate 30. On cultivated land with vegetation cover, fields will appear reddish on Infrared Ektachrome photos. On black-and-white infrared space photos, confusion is possible between cultivated and forest land because of the complex tone values of various vegetation and soil

types (fig. 7–2). Some bare sandy soils are light in tone as are fields of lush vegetation such as alfalfa. Bare fields are generally light in tone on IR–89B photos and tan to white in color on Infrared Ektachrome photos with variation caused by soil differences.

Open Bodies of Water, Rivers, and Canals

Open bodies of water are generally delineated with ease on aerial photography (fig. 7–3) but may be confused with swampy areas and cloud shadows on space photos (Color Plate 30). On Infrared Ektachrome photos water has various tones or colors depending upon impurities (fig. 7–4). Remnant river courses can be seen on aircraft and space photographs as in Color Plate 30. Oxbow lakes appear nearly black because of the clear water in these areas. Generally, on Infrared Ektachrome photos clear water is dark blue to black while turbid water is light blue to tan. Often in forest areas on S065 photos of the Mississippi-Louisiana area, clear water is difficult to see on Infrared Ektachrome photos, where it runs through deciduous forests, because both are dark in color.

On black-and-white infrared photos, open bodies

120

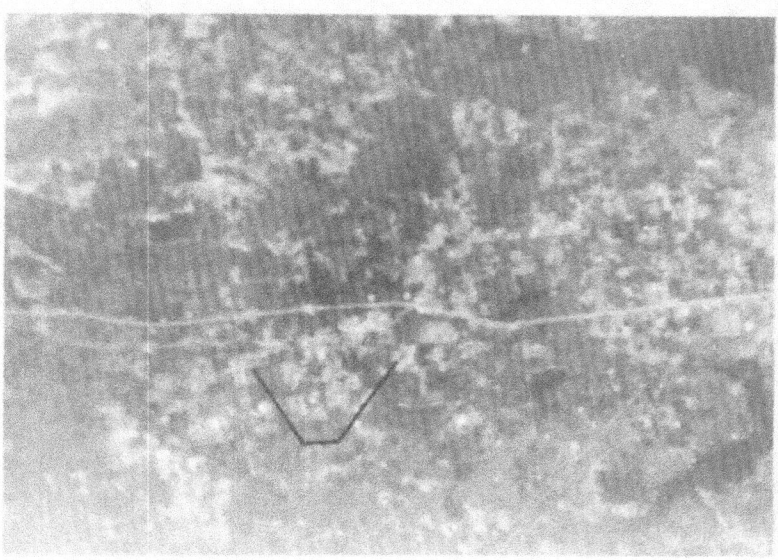

FIGURE 7-1.—Two major classifications of land are visible on the oblique photograph (*right*) above—timbered and nontimbered. This photograph covers the area indicated by the caret on the enlargement of the Apollo 9 Pan-25A image (*left*). Nontimbered areas are largely bare ground with some pasture or grassland. Pine plantations such as the dark rectangular plot left of center on the oblique photograph may be too small to identify on space photographs, but supporting aircraft photos permit positive identification.

of water are very dark in tone unless sediments are present to increase infrared reflectance. Thus, for mapping watercourses, black-and-white infrared space photos are very useful (fig. 7–2).

Urban and Industrial Areas

Generally, a discontinuity in terrain or vegetation appearance will be a clue to the location of urban areas, but there are many features that can cause confusion. Color photos and black-and-white photos taken with a Wratten 25A (red) filter are very useful in locating urban features because of their color or tone values and high spatial resolution (fig. 7–5).

Convergence of evidence may be required in locating urban features on space photos. The following items often will serve as clues to the presence of urban features. (See fig. 7–6 for some of these features.)

(1) Confluence of roads and railroads

(2) Smoke or haze concentration
(3) Cleared land
(4) Harbors
(5) Large airports
(6) Regular patterns of streets and buildings
(7) Natural terrain features such as mountain passes, flat valleys in mountainous areas, and the mouths of major streams

Major Roads

Major roads are recognized because of their linear appearance. It is possible to confuse roads with canals on space photos because some canals are filled with silted water and are similar in appearance to roads on many types of photos. On black-and-white infrared photos, most canals and rivers are dark in tone while roads are light. Some confusion can be eliminated by comparing several spectral bands. (See Color Plate 30 and figs. 7–2 and 7–5.)

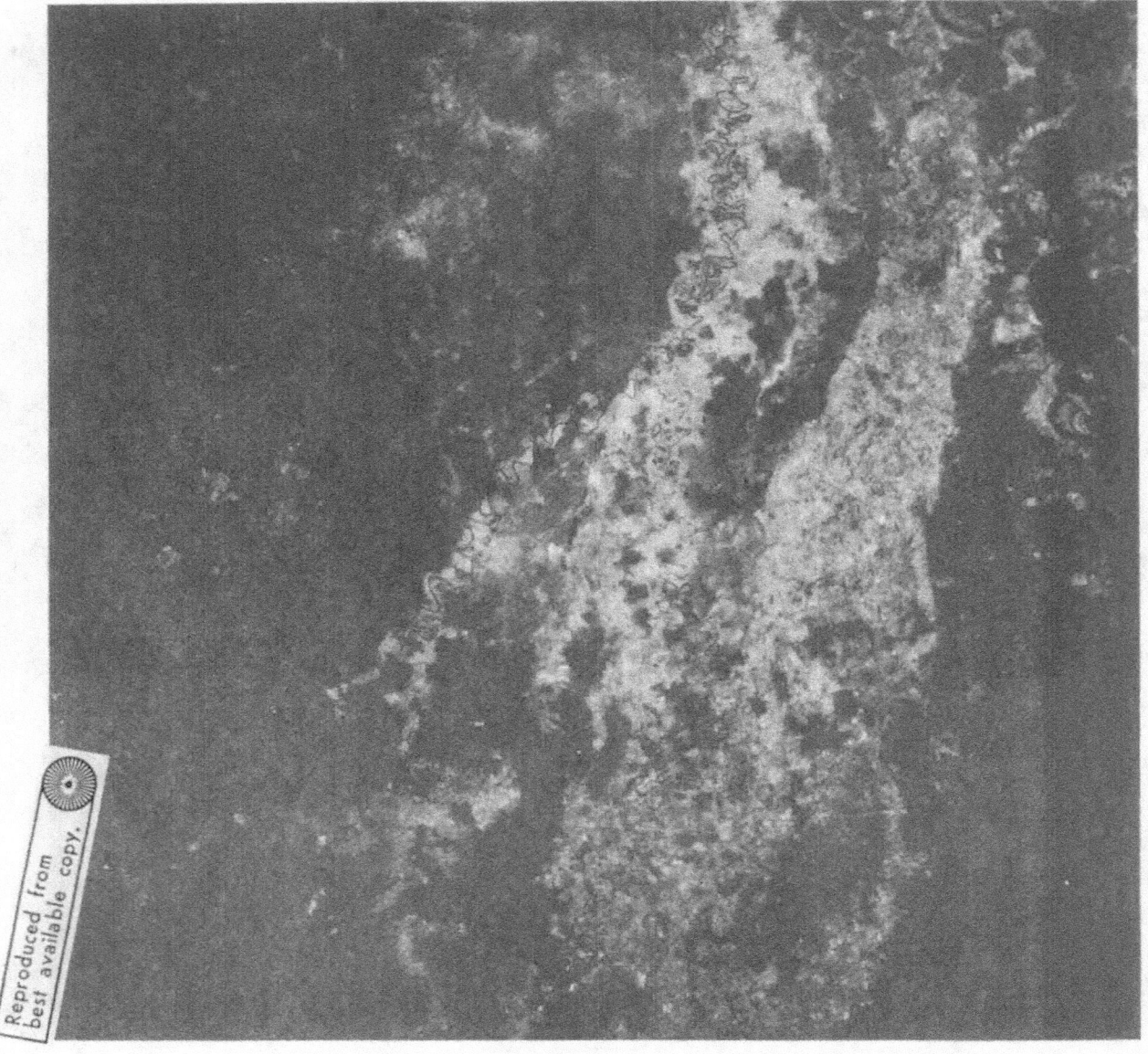

FIGURE 7–2.—Black-and-white IR–89B space photos, such as this one taken by the Apollo 9 astronauts, are very useful for delineating watercourses and open-water bodies because of the nearly complete absorption of infrared radiation by clear water. Spatial resolution and complex tone values make vegetation classification a difficult task on space photos, however. Compare with the left half of Color Plate 30.

SPACE PHOTOINTERPRETATION TEST OF THE MISSISSIPPI-LOUISIANA AREA

A test has been performed utilizing recognition techniques developed in this study to determine the usefulness of space photography for classifying land areas in the Mississippi-Louisiana area covered by S065 photographs. The objective of the test was to determine how consistently each of the land and vegetation classes previously described could be identified on Infrared Ektachrome space photography. The test categories discussed previously in this chapter were chosen (deciduous forests, pine forests, etc.).

122

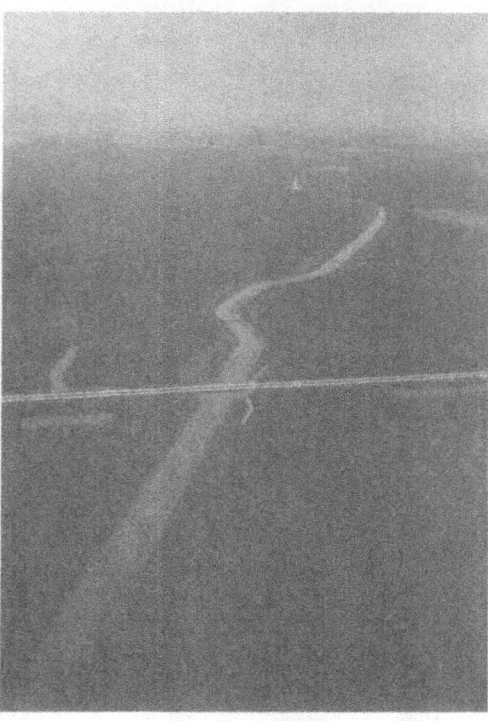

FIGURE 7–3.—Canals for irrigation and drainage are readily seen on the S065 Pan-25A photo at left and on Color Plate 30 at G. Note how the color or tone of linear features is important in separating roads from canals. On space photographs confusion may exist unless care is exercised in evaluating relationships to other features. Low-altitude vertical or oblique photographs such as the one on the right (location on space photograph indicated by caret) can be used to verify the identity of features in question.

FIGURE 7–4.—On this enlarged portion of a Pan-25A space photograph taken by the Apollo 9 astronauts, rivers such as the Mississippi and the Yazoo near Vicksburg, Miss., are clearly seen. Note the high sediment load carried by the Yazoo, which drains agricultural areas east of the Mississippi River as seen at F in Color Plate 30. The town of Vicksburg can be seen by the patterns of streets and buildings where the two rivers meet (point H in Color Plate 30).

124

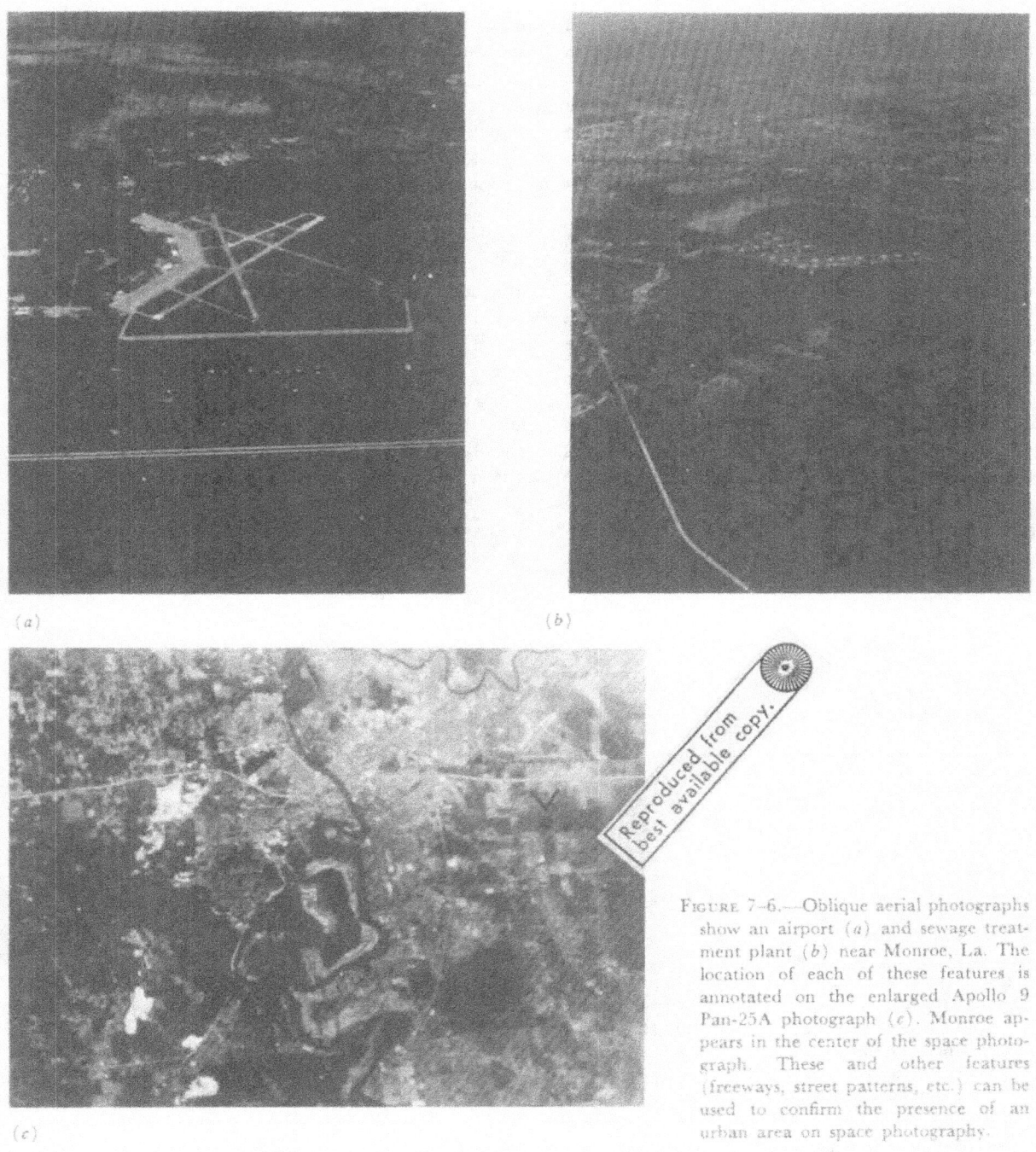

(a)

(b)

(c)

FIGURE 7–6.—Oblique aerial photographs show an airport (a) and sewage treatment plant (b) near Monroe, La. The location of each of these features is annotated on the enlarged Apollo 9 Pan-25A photograph (c). Monroe appears in the center of the space photograph. These and other features (freeways, street patterns, etc.) can be used to confirm the presence of an urban area on space photography.

Interpretation was performed on the two images that constitute Color Plate 30. Each was viewed on an Itek rear-projection viewing screen at 10 × enlargement. A minimum size of feature to be recognized was determined to be 100 acres. A feature of this size could easily be delineated and discriminated from other features when viewed at this scale. Also, this minimum (or possibly a larger one) appears to be workable for land-use mapping on space photos of this quality.

Two types of training material were prepared for use by the interpreters. Each interpreter was instructed to read the descriptive material presented in this section and to study the aerial oblique photographs so that he could become familiar with the characteristics of the ground scene. Then each interpreter was asked to study, on the Itek viewing screen, a number of training samples (25 on each frame), which were selected to represent the variety of tonal and textural characteristics for each of the categories.

Ten interpreters (all of whom had experience with tests similar to this one) were trained in the manner just described. Each was then asked to examine a total of 39 outlined areas (100-acre minimum) and to classify them as belonging to one of the test categories. These test examples were chosen to represent, as far as possible, the range of variability exhibited by the categories on the photographs. The identification of each outlined area was confirmed by locating it on either the low-level aerial oblique or vertical photographs obtained shortly after the Apollo 9 mission. There is sufficient detail (tree-crown characteristics, field pattern, etc.) on these Ektachrome and Infrared Ektachrome photographs to permit the broad categories to be positively identified.

The results from this test appear in table 7-2. Displayed in that table are the composite results for all 10 interpreters. The reader is invited to compare the results by interpreters (data from left to right) with actual ground truth (data down the columns). For example, a total of 60 actual deciduous forest plots were used in this test (total for first column under D). Of this number, 49 were identified correctly by the interpreters (italicized number in first column). The percent correct for this category is 81 percent ($49/60 \times 100$). Similarly, the italicized numbers that occur along the diagonal contain the total number of *correct* identifications for each category. These results are expressed as percent values at the bottom of the table. Other entries in the table represent *commission* errors in which a feature from a given cate-

TABLE 7–2.—*Summary of Photointerpretation Test Results, Mississippi-Louisiana Area*

Photointerpreter's results	Ground truth										Total fields seen by photointerpreter	Commission error
	D	P	M	Cv	Cb	Cf	Wo	Wr	R	U		
D	*49*		7				2				58	9
P	1	*27*	13	4			1				46	19
M	5	2	*28*				1				36	8
Cv				*74*						5	79	5
Cb		1		2	*20*					3	26	6
Cf						*18*				1	19	1
Wo	1						*36*			1	38	2
Wr								*18*	6		24	6
R								2	*34*		36	2
U	4		2			2				*20*	28	8
Total items	60	30	50	80	20	20	40	20	40	30	390	
Number incorrect	11	3	22	6	0	2	4	2	6	10		66
Percent correct	81	90	56	92	100	90	90	90	85	67		
Percent commission	15	41	22	6	23	5	5	25	5	28		

NOTE.—Numbers in table indicate the cumulative number of areas identified by 10 interpreters. Numbers in italic indicate the number of areas identified correctly. Symbols used are identified as follows:

D Deciduous forest
P Pine forest
M Mixed deciduous and pine forest
  Cultivated land:
Cv   Vegetation covered ground
Cb   Bare ground

Cf Fallow ground
Wo Open bodies of water (lakes, reservoirs, etc.)
Wr Rivers and canals
R Roads
U Urban and industrial areas

gory was incorrectly identified as belonging to another category. For example, the number 1 (first column, second entry) indicates that in one case a deciduous forest stand was incorrectly identified as a pine forest stand. The types of *omission* errors also are determinable from a study of entries in the vertical columns.

The percent commission error is summarized for each category. These percentages reflect how often other categories were confused with a given category. The calculation is given as follows:

$$\frac{\text{Total seen by photointerpreter} - \text{total correct}}{\text{Total seen by photointerpreter}} \times 100$$

For example, a total of 58 test items were called deciduous forests by the photointerpreters. Of these, 49 were correctly identified. Thus, in nine cases, items in other categories were incorrectly identified as deciduous forest. The percent commission is

$$\frac{58 - 49}{58} \times 100 \doteq 15$$

The following statements can be made regarding the results in table 7–2:

(1) Although interpreters can recognize pine stands 90 percent of the time, they have some difficulty in identifying mixed and deciduous forests on imagery taken at this time of year. Also, the commission error for pine is high (41 percent) indicating that other forest types are often confused with pure pine stands.

(2) If all three categories of forest cover are grouped in a single category, it becomes possible to separate forested land 74 percent of the time from all other categories, and the commission error is only 6 percent.

(3) Three categories of cultivated fields can be distinguished at least 90 percent of the time.

(4) Open bodies of water and linear features (roads and rivers and canals) are consistently identifiable (90, 90, and 85 percent, respectively), especially if there is sufficient contrast between the linear feature and its background.

Thus it seems that, given the success with which these categories can be recognized, it is possible to conclude that broad land-use maps can be accurately prepared from space photography. Of course,

complete correlation does not exist between identifying individual features in this test and mapping an entire area on a space photo. However, with the area minimum established here and the manner in which the test examples were chosen to represent the variety of image characteristics, it seems reasonable to conclude that broad type mapping could be successfully accomplished on Apollo 9 space photographs.

SEQUENTIAL ASPECTS

The value of the time dimension for vegetation inventories using conventional aerial photography has been recognized for some time. Two major benefits can be derived by interpreting photographs taken on different dates:

(1) Knowledge of the phenological patterns of the vegetation being studied permits specification of time(s) of year when plant species can be most easily separated one from another.

(2) When repetitive cover is obtained for several years, changes in the vegetative cover can be monitored. The effects of manipulation (logging, clearing, burning, etc.) can be measured and treatments can be evaluated.

Low-level aerial oblique photographs were obtained on two different dates during 1969 for comparison purposes. Shortly after the Apollo 9 space photographs were obtained, the first of these low-level missions was undertaken to obtain additional information about the vegetation at the time when the deciduous hardwoods were not in leaf. This information was used to aid in the interpretation of the space photography, and has been mentioned earlier in the chapter. In November 1969, the area was revisited to document the appearance of the forest cover during the fall color change (when the deciduous trees lose their leaves). An example of the photography obtained on these two dates for a portion of the area appears in Color Plate 32.

Comparison of the oblique photograph taken on March 24, 1969, with the oblique photograph taken on November 11, 1969, supports the conclusion that hardwood and conifer forest types can best be distinguished on winter or early spring photography. The dark-red color of conifer stands contrasts sharply with the blue-gray color of deciduous trees as seen on Infrared Ektachrome pho ographs taken at that time of year. This distinction is not nearly so evident during the fall. Of course, a particular hardwood species might have a unique signature during the fall as a result of its color change which would aid in its identification, but the overall value of winter photography for distinguishing between two major classes of vegetation (as is the objective with Apollo 9 photointerpretation) is quite clear.

Identification of the mixed-deciduous and pine-forest category is also best made at this time because of the difference in infrared reflectance of the two components. The difficulty in consistently recognizing this category on Apollo 9 photographs lies in the resolution limitation. Patches of hardwood and conifer trees are often difficult to resolve on space photography, and a mixed stand is often categorized as a deciduous forest stand. The dark-red color of the pine trees is frequently not detectable, and the stand is incorrectly identified. This is merely a problem of resolution, however, and the sequential aspects cannot be used to improve identification.

8

Multistage Sampling of Earth Resources With Aerial and Space Photography

PHILIP G. LANGLEY

Surveying Earth resources from orbiting satellites fitted with remote sensors promises to be one of the significant technological developments of our time. The possibility of obtaining national, continental, or even global coverage with standard specifications as to altitude, sensors, and time over a particular latitude offers a unique basis for developing integrated Earth resources survey systems. Information from these surveys would be integrated with computer-oriented resource information systems that would provide the means to assess the quantity, quality, and location of Earth resources at any given time. The information would be used, for example, to inventory natural resources, monitor crop development, locate new resource possibilities, and establish optimal distribution patterns.

The cataloging of Earth resource information on a vast scale, however, is not a simple matter. One must consider the desirable levels of accuracy and precision of the information in terms of economic feasibility. Obtaining detailed and complete data for every parcel of land over vast areas is not now feasible for several reasons. First, Earth resource data obtained from remote sensors, particularly from orbital altitudes, do not correlate perfectly with ground conditions; thus, errors are introduced into the system. Second, certain detailed resource data can be obtained only on the ground, for example, bole and growth measurements on forest trees or bushels of wheat per unit of land area. Third, even if complete data were available, not enough computers or storage banks are available to handle this information easily. Therefore, the only feasible method of obtaining detailed resource information, applicable to large land areas, is by means of sample estimates—even when remote sensors are used from space.

In forest surveys, aerial photographs are presently being used to improve sampling efficiency (Aldrich, 1968). Stratified sampling, in which the strata are defined by delineating areas of relatively homogeneous forest types on photographs, has been one of the widely used techniques. Double sampling, with stratification at the first phase, has been another; and double sampling with regression is still another useful technique. These forest sampling techniques are usually geared to incorporate medium-scale resource photography generally available from public agencies.

The development of high-altitude aircraft and spacecraft has increased interest in the possibilities of applying small-scale photography to forest inventories—particularly in those surveys that are desired at frequent intervals. There is no "standard" sampling procedure applicable to all resource inventories using space photographs. The design used in a particular situation should take into account the kinds of population parameters being estimated, the distribution of the population variables used to estimate the parameter, the existing information relating to these variables, and the optimum allocation of funds available for the survey. In forest sampling, even a cursory inspection of space photographs strongly suggests that their use in an inventory calls for some form of multistage sample design. The very small scale and relatively low resolution of satellite imagery makes it impractical to correlate the data directly to ground measurements. With this small-scale imagery, one can survey vast land areas rapidly, but it is difficult to identify specific tree species, or even stand composition, except by gross image and physiographic characteristics. Furthermore, sample locations, small enough to be measurable on the ground, cannot be accurately and easily located from the space imagery. But though it is not generally feasible to accurately locate ground plots of small area directly from space photography, it is feasible to correlate larger scale aerial photography to the space photography on a sample basis. Then, still larger scale aerial photography can be located within the previous aerial coverage. This process can be continued until ground plots can be located directly from the last aerial stage. Using large-scale color photography, one can easily locate sample plots on the ground, provided the locality is known; and he can do a reasonably good job of predicting tree species, by size class.

To increase the efficiency of a multistage resource sample survey using space and aerial photography, it is necessary somehow to translate the remote-sensor data to information that relates to the resources of interest in a particular survey. One convenient way of doing this is to utilize multistage sampling with variable probabilities of selection at every stage. In such a technique, the sampling probabilities are formulated from the additional information made available by virtue of the increasingly finer resolution of remote-sensor data at each sample stage. At the last stage, measurements are obtained on the ground. The ground measurements are projected back through the sampling formula to obtain estimates applicable to the entire area of interest.

VARIABLE PROBABILITY SAMPLING THEORY

One-stage variable probability sampling, often referred to as PPS (probability proportional to size) sampling, has been applied in many types of surveys, including agriculture, forest, and census. To understand the potential gain of variable probability sampling over simple sampling with equal probabilities, let us look at the latter. In simple random sampling, we may estimate the total resource quantity on a tract of land as

$$v = \frac{N}{n} \sum^n v_i$$

in which

N = the number of units in the population
n = the number of sample units
v_i = the quantity of the resource measured on the ith sample unit

This formula may be rearranged to

$$v = \frac{1}{n} \sum^n \frac{v_i}{1/N}$$

From sample theory, we know that $1/N$ is equal to the probability of selecting the ith unit at the jth draw when sampling either with or without replacement. Therefore,

$$v = \frac{1}{n} \sum^n \frac{v_i}{p}$$

and each ratio v_i/p estimates the population total. The true variance of this estimator (i.e., not the estimated variance) when sampling with replacement is

$$\text{Var}(v) = \frac{1}{n} \sum^N P \left(\frac{V_i}{P} - V \right)^2$$

in which V is the true resource value and the other terms are as defined earlier. Notice that the variance arises by virtue of the variation among the population units V_i and is unaffected by the constant probability of selection. Contrast this condition to the case of PPS sampling in which the probabilities of selection are defined as

$$P_i = \frac{C_i}{\sum^n} C_i$$

in which C_i is a prediction, obtained by means of remote sensing, of the resource quantity contained in the ith population unit. The estimator may now be written as

$$v = \frac{1}{n} \sum^n \frac{v_i}{p_i}$$

with variance

$$\text{Var}(v) = \frac{1}{n} \sum^N p_i \left(\frac{V_i}{p_i} - V \right)^2$$

Here again, each ratio V_i/P_i estimates the population total. But in this case, the higher the linear correlation between V_i and P_i, the more nearly alike are the ratios over all units and the closer they are to the true population total V. Therefore, the variance is lowered by virtue of the smaller deviations of the ratios V_i/P_i from V. It can easily be shown that perfect proportionality between V_i and P_i over all population units leads to a variance of zero, regardless of sample size. Negative correlation, on the other hand, leads to an extremely high variance—greater than with simple random sampling. Therefore, the development of photointerpretation techniques that yield data highly correlated to larger scale aerial data or ground data is crucial to the successful implementation of this sampling technique.

The estimator v is unbiased as long as the sample units are physically drawn with probability proportional to the prediction. This is easily accomplished by a method known as list sampling. To execute the method, predict the relative amount of resource present in each population unit by means of photointerpretation. List these predictions as shown in table 8–1, column 2. Then make a cumulative list as in column 3. When all population units have been listed, the total (C) of all units will be shown in the last entry of column 3. To select a sample of size 3, say, draw at random with equal probability three numbers between 1 and C. Compare the three random numbers with the cumulative list (column 3) and assign each one to the population unit whose interval includes the number. For example, if one of the numbers drawn at random is 143, unit 5 is included in the sample because 143 is greater than 131, the upper interval bound of unit 4, but less than (or equal to) 222, the upper interval bound of population unit 5. Since the length of each interval is proportional to the predicted resource quantity of the corresponding population unit and the random numbers are drawn with equal probability, the probability of including a population unit in the sample is proportional to its predicted quantity.

The selection process is extended to successive stages by treating each unit included in the primary sample as a subpopulation and repeating the process independently in each primary unit sampled. A two-stage sample estimate of the total resource quantity in an area may be symbolically described as

$$v = \frac{1}{m} \sum_i^m \frac{1}{p_i n_i} \sum_j^{n_i} \frac{v_{ij}}{p_{ij}}$$

in which

m = the number of primary units included in the sample

p_i = the probability of selecting the ith primary unit

n_i = the number of observations included in the ith primary unit

v_{ij} = the resource quantity measured in the jth subunit of the ith primary unit

p_{ij} = the conditional probability of selecting the jth subunit given the ith primary unit has been selected

Irrespective of the number of sample stages, an unbiased estimate of the sample variance may be calculated from the first-stage estimates by using the equation

$$\text{var}(v) = \frac{1}{m(m-1)} \left(\sum_i^m \frac{v_i^2}{p_i^2} - mv^2 \right) \qquad (1)$$

Hence, the multistage estimates are obtained for each primary sample unit, then these are entered

TABLE 8–1.—*Hypothetical Data Illustrating the Mechanics of List Sampling*

Population unit size	Predicted resource quantity	Cumulative prediction	Probability of selection, P_i
1	25	25	25 /1000 = 0 .025
2	38	63	38 /1000 = .038
3	63	126	63 /1000 = .063
4	5	131	5 /1000 = .005
5	91	[a] 222	91 /1000 = .091
6	15	237	15 /1000 = .015
.	.	.	.
.	.	.	.
.	.	.	.
N	42	[b] 1000 (C)	42 /1000 = .042

[a] Unit 5 is included in the sample as a result of drawing the uniform random number 143.

[b] C = total of cumulative predictions.

into equation (1) to estimate the sample variance from which the estimated sampling error is computed.

SPACE PHOTOGRAPHS

There are many ways in which the basic multistage variable-probability sampling design can be applied to a resource inventory using space and aircraft photography. If a large area is being covered, a sample of space photographs may be drawn on while further subsampling is carried out. In variable-probability sampling, the entire population of space photographs must be examined to determine the selection probabilities of the primary units. If this is not feasible, a simple random sample of space photos may be drawn with equal probability provided the spatial distribution of the population is fairly uniform or random, but not clustered. If extensive clustering is present, stratified or systematic cluster sampling should be considered. The probability of selecting the cluster may be proportional to the total number of images included in each cluster.

After the space photos are selected, they are prepared for subsampling. Subunits may be laid out in the form of a grid as we did in a forest inventory using Apollo 9 Infrared Ektachrome photographs covering 5 million acres in Louisiana, Mississippi, and Arkansas (Langley et al., 1969). In that survey, we partitioned the space photographs into 4- by 4-mi squares (fig. 8–1). Regardless of how the space photograph is partitioned, the subunits should be identifiable from an aircraft during the subsampling phase. More important, they should be of a size that variation between units will be small or controllable by means of photointerpretation. The variation within units is controlled by appropriate subsampling and photointerpretation.

Stratification also should be considered if the forest characteristics differ among areas that exhibit the same image characteristics from one area to another. For instance, in the Mississippi River Valley area, upland pine and hardwood apparently have different volumes per unit of forest land than either upland or bottomland hardwood alone. By stratifying out the pine areas we reduced our sampling error by one-third. In the Mississippi Valley survey we selected a sample of five 4- by 4-mi squares from the Apollo 9 photographs. Two squares were drawn from the pine stratum and three from the hardwood stratum.

Since we worked with only two Apollo 9 frames in the Mississippi Valley survey, the annotated 4- by 4-mi squares constituted the population from which our primary sampling units were selected. With units this large (10 240 acres), we were able to achieve three objectives: (1) the units were readily identifiable from an aircraft for rephotographing at a larger scale, (2) they were large enough so that a meaningful prediction could be made as to the relative amount of timber volume contained in each unit, and (3) they were large enough that the between-unit variation in average timber volume per acre of forest land would be relatively low within strata. The success of the study depended on meeting these three objectives.

After delineating the population units on the space photos, each square was examined through a Bausch & Lomb Zoom 70 stereoscope under 7.5 $\times$ magnification (fig. 8–2). From this examination, the interpreter estimated the proportion of each square occupied by forest land (fig. 8–3). This proportion was used to predict the relative timber volume in the square. The contention was that, on areas of this size, forest area was proportional to timber volume.

After predictions were made on all the squares, a primary sample was drawn at random with probability proportional to the prediction.

SUPPORT AERIAL PHOTOGRAPHS

Between April 15 and April 24, 1969, aerial photography of each primary sampling unit selected for the inventory was obtained. The camera package consisted of a Crown Graphic camera with a Polaroid back and two Mauer KB–8 70-mm cameras mounted in a single frame (fig. 8–4).

The first aircraft imagery obtained over the primary sample units consisted of 1/60 000-scale photographs taken on Polaroid film through a Wratten-15 filter (fig. 8–5). Polaroid was used so that we could obtain and interpret the imagery while still airborne. The scale was chosen so that the entire sample square would be covered by the 5-in. film format in one pass. After obtaining the Polaroid photography but while still airborne, we

133

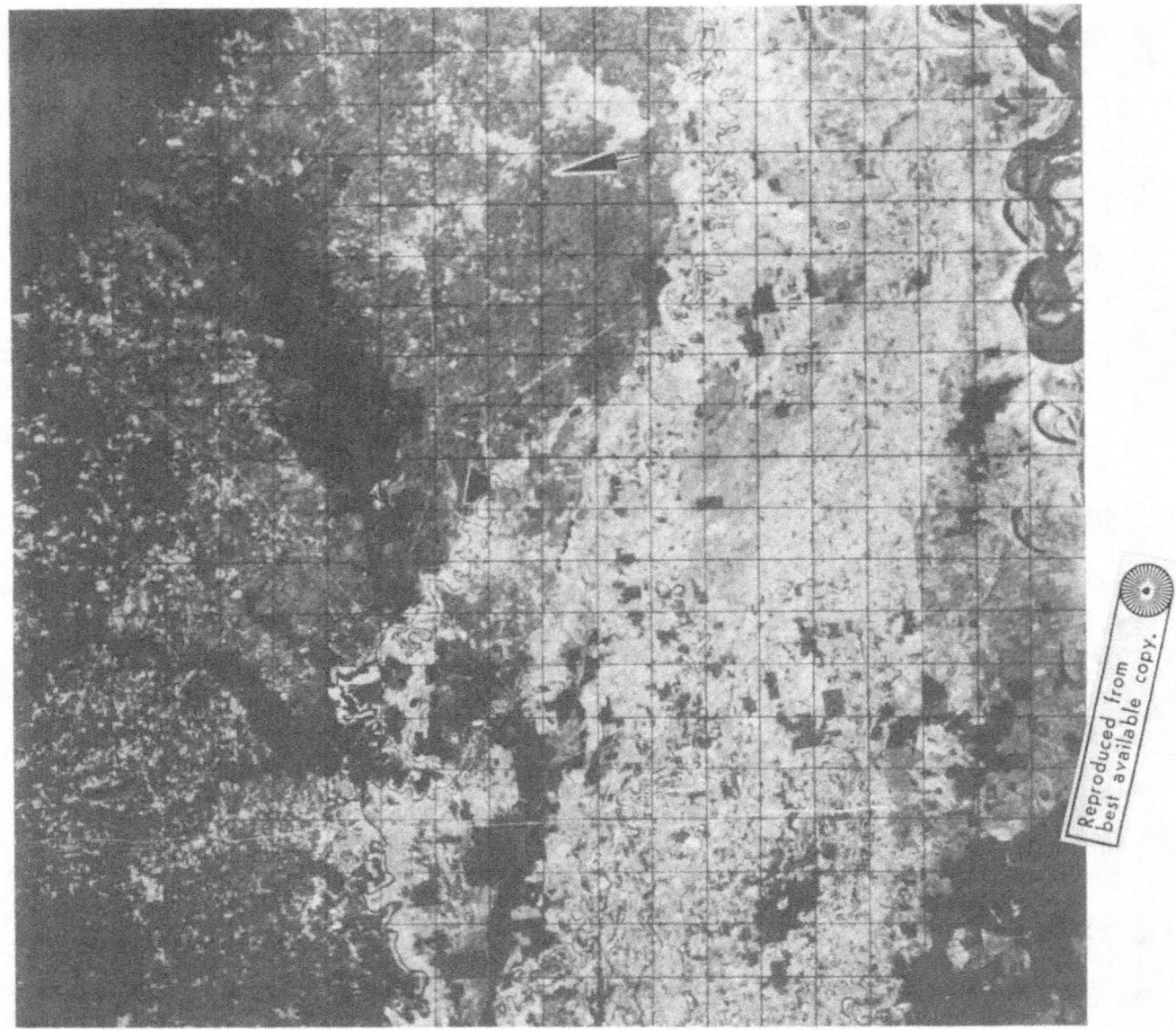

FIGURE 8–1.—A black-and-white reproduction made from Apollo 9 Infrared Ektachrome 3740 enlarged approximately three times. The grid of 400 4- by 4-mi squares was used to predict timber volumes for the first level of inventory information. An Infrared Ektachrome photograph of this area may be seen in Color Plate 30. The arrow indicates one of the 4- by 4-mi cells selected for subsampling.

prepared an aerial mosaic of the area and super-imposed a strip grid partitioning the entire 4- by 4-mi square into subsampling units (fig. 8–6). The strip subunits were of such a dimension that selected ones could be rephotographed in one pass at a scale of 1/12 000 with 70-mm photography.

Immediately after the Polaroid mosaic was prepared, each strip was examined and a new prediction was made as to the relative quantity of timber contained in each. The forest information contained on the 1/60 000-scale photos was sufficiently detailed to allow the prediction of timber volumes

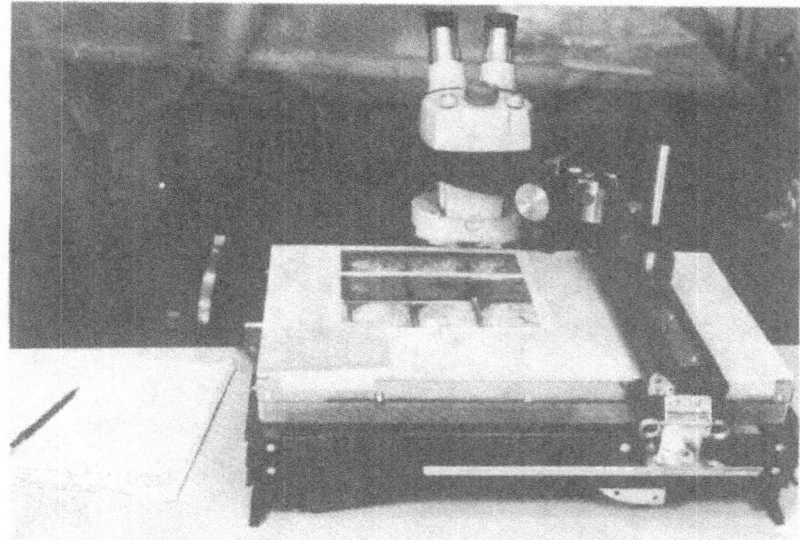

FIGURE 8-2.—Apollo 9 Infrared Ektachrome photographs were interpreted with the aid of a Bausch & Lomb Zoom 70 stereoscope with 7.5 × magnification. Apollo 9 Infrared Aerographic (Wratten 89B filter) and Panchromatic (Wratten 25A filter) films were also used to aid in separating forest from other land uses.

needed in selecting a sample of strips for the following stage. At the same time, the photos were not so detailed as to hinder the interpreter in his rapid appraisal of the imagery. The new information was used to determine sampling probabilities by which two strips were selected in each square. A hand-operated adding machine and a table of random numbers were all that were needed to make the selections with probability proportional to our predicted timber volumes by means of list sampling.

The selected strips were marked on the Polaroid mosaic, which was then used by the pilot as a flight map for the next stage. The two selected strips were rephotographed with the two 70-mm cameras. One camera, equipped with a 1½-in. lens, covered the entire strip at a scale of 1/12 000. The other camera, equipped with a 9-in. lens, provided a sample of 1/2000-scale color photographs simultaneously (fig. 8-7). The 1/2000-scale photos consisted of a systematic sample with a random start taken down the center of the strip.

The white square in figure 8-6 indicates the position of one of the 1/12 000-scale 70-mm photographs (fig. 8-8). The black square on the 1/12 000-scale photograph indicates the location of a sample 1/2000-scale color photograph (fig. 8-9).

Back in the laboratory, the strip boundaries were delineated on a mosaic of the 1/12 000-scale photographs. Then, the photo coordinates of these boundaries were digitized at 0.01-in. intervals using a Bendix datagrid digitizer (fig. 8-10). From these data, the strip areas were computed. The proportion of the strip covered by the 1/2000-scale photographs was computed from the number of 1/2000-scale photographs in a strip and the area of the strip. The inverse of this proportion was used to expand the timber-volume estimates from the cluster to the strip level.

The 1/2000-scale clusters of color photographs constituted stage 3 in our survey design. They were obtained in triplets so as to provide stereoscopic coverage of the center frame. The number of triplets ranged from 13 to 20 per strip.

The center photograph of each triplet was partitioned into four square plots, each plot being about 0.6 acre in the Mississippi Valley area (fig. 8-9). Plots of this size are convenient for locating and measuring on the ground.

GROUND PLOTS

Each of the large-scale photoplots was examined stereoscopically. Estimates were made as to stand height, determined by means of a parallax wedge described by Wert and Myhre (1967); crown coverage, determined by counting dots on a grid; and crown diameter. From these estimates, timber volumes were predicted on all plots by using an

FIGURE 8–3.—Each 4- by 4-mi square within the Apollo 9 frame was examined, and the proportion of the area occupied by forest land was estimated. The center square is the one indicated by the arrow in figure 8–1 and was one of those selected for subsampling.

FIGURE 8–4.—This aerial camera setup was used to obtain aerial photography of primary sampling units selected from the space photograph. Included are (1) a Crown Graphic with Polaroid back, (2) a J. A. Mauer KB–8 70-mm camera with a 1½-in. lens, and (3) a J. A. Mauer KB–8 with a 9-in. lens.

FIGURE 8–5.—Polaroid photographs (1/60 000) were taken over each primary sampling unit
selected for the first stage in the multistage sample. The mosaic shown is for the 4- by
4-mi square outlined in white in figure 8–3.

aerial photovolume table (Avery, 1958). At this resolution level, it was easy to eliminate from further consideration those plots that fell on nonforest land or that contained no timber volume. From this information, one plot per strip was selected for measurement on the ground. Again, the selection probabilities were proportional to predicted timber volumes at that level.

For each plot drawn, a packet of photographs was made up containing each scale of photography covering the plot. This packet was given to the field crew to aid in locating the plots on the ground. Little difficulty was encountered in locating the field plots.

After a field plot was located and laid out on the ground, each tree's diameter was measured and recorded along with the species. The diameters were used to predict each tree's volume from a volume table but modified by the timber cruiser's ability to adjust for defects and deformities in the tree. By using these predictions, four to six trees were selected on which bole measurements were obtained by means of a precision optical dendrometer (fig. 8–11). The wood volumes of these sample

FIGURE 8-6.—The 4- by 4-mi primary sample covered by the Polaroid mosaic was divided into subsampling units using a transparent strip grid. The area outlined in white indicates the position of the 1/12 000-scale photograph shown in figure 8-8.

trees were later computed from the dendrometer measurements using a computer program by Grosenbaugh (1964).

SAMPLING FORMULA

To obtain timber volume estimates applicable to the entire survey area, the measured tree volumes were expanded back through the sampling formula. In the Apollo 9 study, the estimated timber volume in each stratum was

$$v = \frac{1}{m} \sum_{i}^{m} \frac{1}{p_i n_i} \sum_{i}^{n_i} \frac{1}{p_j} \frac{A_j}{a_c} \frac{1}{p_p l_p} \sum_{i}^{t_p} \frac{v_k}{p_k}$$

in which

$m =$ the number of 4- by 4-mi squares included in the primary sample

$p_i =$ the probability of selecting the ith sample square

$n_i =$ the number of sample strips in the ith 4- by 4-mi square

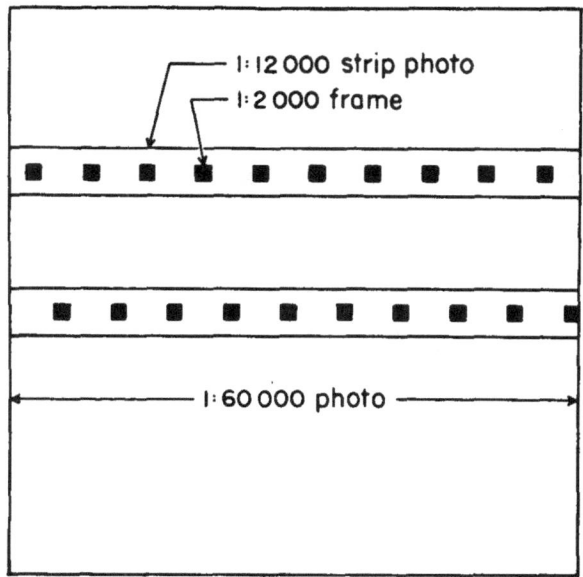

FIGURE 8–7.—The scaled diagram shows how the two 1/12 000-scale 70-mm sample strips and 1/2000-scale 70-mm color samples are related to each other and to the 1/60 000-scale Polaroid photograph.

p_j = the probability of selecting the jth sample strip in a sample 4- by 4-mi square area

A_j = the total area of the jth sample strip

a_c = the area covered by the cluster of 1/2000-scale 70-mm photographs within a strip

p_p = the probability of selecting the pth plot from the cluster of plots delineated on the 1/2000-scale 70-mm photos in a strip

t_p = the number of sample trees measured on the pth plot

v_k = the measured volume of the kth sample tree on a selected ground plot

and

p_k = the probability of selecting the kth sample tree

RESULTS

The best results were obtained on the 5 million acres covered by the Apollo 9 photographs in Louisiana, Mississippi, and Arkansas. With only 10 ground plots, constituting a sampling fraction of one to a million in terms of area, we obtained an estimate of 2.225 billion gross ft³ of timber, with an estimated sampling error of only 13.0 percent. Half of this error was attributable to

inadequate tree-volume tables we used on the ground to correlate with the dendrometer measurements.

The gain in the sampling error we achieved by using information from the space photos was quite substantial. If we had used the same sampling plan—but with equal probabilities at the first stage and without the benefit of stratification—we would have incurred a sampling error of 30.7 percent. Stratifying on the space photos brought the error down to 22.5 percent. Then, by using probability sampling in selecting the primary sample, the error came down to 13.0 percent. Hence, a 58-percent reduction in the sampling error (from 30.7 to 13.0) was directly attributable to the information gleaned from the Apollo 9 photos.

In a similar survey conducted in Georgia, we were less fortunate. There, we were unable to show a gain in the sampling error as a result of information on the space photos. This was due to low correlation between our predicted timber volumes on the primary sample units and our estimated volumes on corresponding units obtained by subsampling. However, as ground data become available, one can develop improved photo-interpretation techniques for use in the next survey.

DISCUSSION AND A LOOK TO THE ERTS

Although we may not have formulated the optimal sample design for the Apollo 9 study, we felt we successfully perceived and dealt with the majority of the problems that might have arisen operationally as well as statistically. As techniques are developed for extracting better information from multispectral data, by either manual or automatic methods, we will be able to inventory vast forested areas better from space and do it rapidly with a minimum of ground work.

Looking ahead to the time when the Earth Resources Technology Satellite (ERTS) will be launched, we have suggested the following modus operandi for a first forestry and agriculture resource information system based on remote sensing from space:

(1) Multispectral data telemetered to Earth from the satellite is processed by computer and immediately converted into condensed tables of predictions relating to resource variables of inter-

FIGURE 8–8.—A 1/12 000-scale photograph of the area outlined by the white square in figure 8–6. The area outlined in black corresponds to the coverage of the 1/2000-scale photograph in figure 8–9.

est to national and local management planners. These predictions could be associated with land units of some arbitrary, but not necessarily equal, size. This information base is continually updated as the satellite obtains new data over a given area.

(2) When specific information is needed about certain resources, a decision is made as to the allowable error of the estimates to be obtained.

(3) Using the information stored in the system, a sample of primary units is drawn according to an appropriate set of selection probabilities. The sample size is determined by the allowable error and the quality of the information currently in the system that relates to the variables of interest. The information on some variables will be more reliable than on others.

(4) An aircraft bearing a full complement of remote sensors, or simply cameras, is dispatched to the sample areas; and new, higher resolution data are obtained. This could be done in one or more stages. If more than one stage is needed, an on-board image analyzer and computer could be programed to determine the subsampling probabilities from the imagery just obtained and to select the sample for the next stage.

(5) At the last aerial stage, a small sample of ground plots is selected for ground measurement.

(6) The ground observations are entered into the sampling formula to obtain the needed estimates applicable to the area of interest.

(7) These final estimates serve to improve the statistical models used in assigning the original re-

140

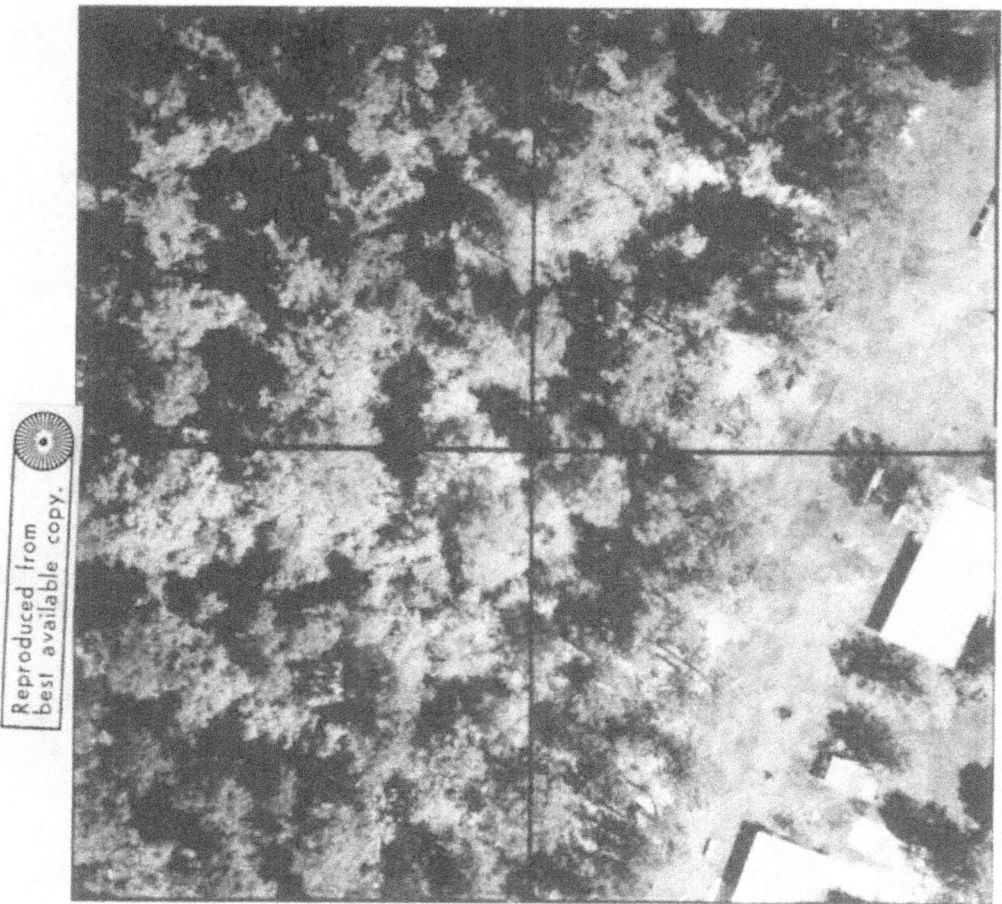

FIGURE 8–9.—A 1/2000-scale photograph corresponding to the area outlined in black on the 1/12 000-scale photograph in figure 8–8. The grid divides the photograph into four sample plots approximately 0.625 acre in size. .

source predictions from satellite data. At the same time, the data base, relating to the parameter for which new estimates have just been obtained, is improved.

Taking this approach, a workable resource information system for forestry and agriculture could be made ready to utilize information immediately upon its receipt from Earth-orbiting satellites and begin supplying needed answers to questions concerning the status of the resource base. In time, the data base would improve in precision to the point where a minimum of aircraft and ground data would be needed to answer specific questions asked by management planners and policymakers. At that point, the system could be queried and answers returned almost instantaneously as to the

quantity and distribution of agriculture and forest resources in any part of the country.

Finally, there are other advantages to an approach of this kind to a resource information system. One advantage is that information about specific variables is improved as the need arises. Hence, an exact definition of the data base is not essential in the beginning. The data base would be improved as the system was used operationally. Another feature is that various users would have the opportunity to participate at a level of sophistication that suited them. They could obtain aerial subsamples using multichannel scanners or simple cameras. Furthermore, they could carry out their own fieldwork and handle their own data processing for the aerial and ground stages of their

FIGURE 8–10.—Photo coordinates of sample strip boundaries outlined on the 1/12 000-scale photomosaic were digitized at 0.01-in. intervals using this Bendix datagrid digitizer.

FIGURE 8–11.—An optical dendrometer was used to make bole measurements on four to six trees on each ground plot.

surveys. These qualities should appeal to other countries wishing to share in the NASA Earth Resources Technology Satellite programs.

SELECTED LITERATURE

ALDRICH, R. C. 1968. Remote Sensing and the Forest Survey—Present Applications, Research, and a Look at the Future. Proc. 5th Symp. on Remote Sensing of Environ., Univ. of Michigan 968: 357–372.

AVERY, G. 1958. Composite Aerial Volume Table for Southern Pine and Hardwoods. J. Forest. 56: 741–745.

GROSENBAUGH, L. R. 1964. STX-Fortran IV Program for Estimates of Tree Populations From 3P Sample-Tree-Measurements. U.S. Pac. Southwest Forest Range Exp. Sta. Res. Pap. PSW–13. Berkeley, Calif.

LANGLEY, P. G.; R. C. ALDRICH; AND R. C. HELLER. 1969. Multi-Stage Sampling of Forest Resources by Using Space Photography—An Apollo 9 Case Study. Vol. 2: Agriculture, Forestry, and Sensor Studies. Proc. 2d Annual Earth Resources Aircraft Program Rev., pp. 19–1 to 19–21. NASA MSC, Houston, Tex.

WERT, S. L.; AND R. J. MYHRE. 1967. Wedge Measures Parallax Separations on Large Scale 70-mm Aerial Photographs. U.S. Pac. Southwest Forest Range Exp. Sta. Res. Pap. PSW–142. Berkeley, Calif.

9

The Use of Small-Scale Aerial Photography in a Regional Agricultural Survey

William C. Draeger and A. S. Benson

Maricopa County, Ariz. (Color Plate 2), was the site of extensive NASA-sponsored research designed to investigate the potential usefulness of small-scale aerial and space photography in the inventory and evaluation of agricultural crops. (See ch. 3 for detailed analysis of small-scale photographs.) Early in these investigations it became apparent that in order to assess fully the operational value of such photography, a regional approach to the research would be necessary.

One of the primary advantages of using small-scale aerial or space photography is that it affords a synoptic view of the Earth's surface (i.e., large areas of land can be seen in their entirety on one or a very few images), suggesting a particular potential usefulness for conducting broad regional resource analyses. Furthermore, few actual inventories as presently undertaken limit themselves to a small area; they are usually geared to large managerial or policy-formulation units such as entire watersheds, counties, or States. Thus, most remote-sensing surveys, when performed operationally, would probably also be geared to fairly large areas so as to provide maximum utility to the ultimate user. Finally, although the development of remote-sensing techniques on small test sites is often quite useful, especially in the early experimental stage, findings of limited tests often cannot be directly applied to the larger operational case. Moreover, the obvious problems stemming from increased interpreter fatigue and data-handling requirements must be acknowledged when large areas are surveyed. The phenomenon of environmental variability also becomes a major factor to be dealt with in the design of information-extraction techniques.

For these reasons, it seemed that one of the most meaningful experiments that could be per-formed with the small-scale aerial photographs would be to make an agricultural survey for Maricopa County as a whole. By so doing, an attempt could be made to answer questions that would arise in such a semioperational survey and that must be solved before the full benefits from the use of high-altitude or space photography can be realized. In addition, it was hoped that such a study might provide clues to the procedures to follow in evaluating synoptic imagery that will become available from the Earth Resources Technology Satellites, ERTS–A and ERTS–B, to be launched in early 1972 and 1973, respectively, and the manned Sky Laboratory (Skylab).

While any of the varied resources of Maricopa County could be the subject of such a survey, none is more important or more amenable to the application of remote-sensing techniques than the agricultural crops. According to recent records, over 10 percent of the land in Maricopa County is under cultivation. The county provides roughly half of Arizona's agricultural crop production and ranks third among all U.S. counties in gross value of such products. Many of the crops grown contribute directly to the livestock- and cattle-feeding industry, in which Arizona ranks eighth nationally. The nature of agricultural cropland makes it especially well suited to such a study. By and large, such land consists of discrete fields, each of which contains a fairly uniform crop that may vary quite rapidly in its phenological characteristics through a seasonal cycle. This characteristic presents an excellent opportunity for the development of techniques that could be quite valuable in their own right and that could contribute to methods applicable to more variable wildland vegetation types. Finally, a very real need exists for inexpensive, accurate, and up-to-date inventories of agricultural crops, as is evidenced by the extensive program carried out by the Statistical Reporting Service of the U.S. Department of Agriculture in cooperation with various State and county organizations.

PRELIMINARY TESTS

As has been described earlier (ch. 3), numerous photointerpretation tests were conducted on a 16-sq-mi area within Maricopa County. These tests were intended to determine the relative value of small-scale aerial photography and Apollo 9 space photography for the inventory of crops and to evaluate the usefulness of multidate and multiband photography for these surveys.

The test results suggested that, for agricultural surveys in the area under study, no significant differences in accuracy of crop identification resulted from the use of Apollo 9 and high-altitude aerial photographs. In addition, they emphasized the importance of the date of photography for the accurate inventory of particular crops and the need to establish the seasonal development of crops in a region before specifying the optimum dates for obtaining photography.

Following these tests, the decision was made to perform a semioperational survey for barley and wheat. This decision was based on the following factors: (1) small grains (of which barley and wheat are the major varieties in Maricopa County) account for approximately 20 percent of the crop acreage in Maricopa County and are important crops for which agricultural statistics are currently prepared using conventional techniques (mail questionnaires and personal interviews with growers); (2) these crops mature and are harvested within the first half of the calendar year, coincident with the time period for which monthly NASA aircraft missions were scheduled during 1970; (3) our previous results indicated that the highest percentage correct identification of any crop was achieved for barley (90 percent using Infrared Ektachrome photos and 91 percent using Pan-25A) by selecting the appropriate month for conducting the test. Thus it was felt that a survey for barley and wheat would provide the greatest opportunity for initial success using a previously untried technique.

Preliminary tests were conducted to determine the specific date or dates of photography and film-filter combinations optimum for the identification of barley and wheat. The results of these tests indicated that Ektachrome MS (2448) photography taken in the months of May and June 1970 should be used for the semioperational survey.

DEVELOPMENT OF THE SEMIOPERATIONAL SURVEY

The administration of a photointerpretation survey involving an entire county, containing nearly 800 sq mi of agricultural land, presented a number of problems not faced on the 16-sq-mi study area. There were three principal questions: (1) Will a sample provide a satisfactory estimate of crop acreage, or is 100 percent interpretation required? (2) Will stratification lead to a more accurate estimate? (3) How much ground information will be required for interpreter training and for evaluation of the interpretation? In an attempt to answer several questions simultaneously, the agricultural area within the county was delineated into six strata based wholly on their appearance in the Infrared Ektachrome Apollo 9 photo. Thirty-two square plots, 2 mi on a side, were allocated to the six strata on the basis of proportional area. Plot centers were located randomly (fig. 9–1). Maps of each plot showing field boundaries were drawn based on their appearance on earlier high-flight photography, and each plot was visited by a field crew at the time of overflights for the months of April, May, and June, 1970. Information gathered included the category of crop growing in each field, the condition of the crop, the percent of the ground covered by vegetation, crop height, and the direction of rows, if any.

To facilitate access to the information for each of the more than 2500 fields present in the thirty-two 4-sq-mi sample plots (comprising a total of more than 80 000 acres), field data were punched on computer cards. Programs were written that made possible the compilation of data by stratum, cell, crop type, and date; and these programs provided for subdivisions or consolidations of fields over time. Thus data are available for each date of photography and for the sequential changes in crop type and condition through the growing season as well.

Based on a knowledge of the distribution and

FIGURE 9-1.—This black-and-white enlargement of an Apollo 9 space photograph shows the portion of Maricopa County containing agricultural lands for which the semioperational survey was performed. Black squares indicate the location of 4-sq-mi plots selected for ground survey at the time of each NASA overflight.

2 miles

variability of crop acreages, tests were conducted to determine the value of stratification based on gross appearance in space photography, and the possibility of sampling within the agricultural areas to obtain overall crop acreages for the county. Analyses of variance indicated that no significant differences existed between strata in terms of acreages of major field crops, thus indicating that stratification would not improve acreage estimates. In addition, calculations indicated that the acreage distribution of major crops was so variable that for any plot size, extremely large samples would be necessary in order to assure acreage estimates that would satisfy accuracy requirements. For example, to estimate the average of wheat with a standard error of ±10 percent of the total acreage using a plot size of 4 sq mi, a 75-percent sample of the total agricultural area would be necessary.

Thus, it was decided that the most efficient and realistic method of estimating crop acreage would entail a 100-percent photointerpretation of the agricultural areas, with ground data being gathered for the 32 square plots only. In this way photointerpretation results could be compared with the ground conditions on the field plots, and the overall photointerpretation results adjusted as appropriate using standard ratio sampling procedures.

Some problems were encountered in the development of a method for compiling photointerpretation data. First, to make a measure of interpretation accuracy, interpretation findings must be tied to some actual unit of land area. However, the preparation of detailed field boundary maps from small-scale photos by the interpreter, while possible, would constitute an extremely time consuming task. Second, the tabulation of interpretation data on the basis of *numbers of fields* is not necessarily indicative of accuracy of *acreage* estimates, which in most cases is the item of interest to the ultimate user. Third, to evaluate "number of fields" data, the researcher must assign arbitrary weight to "correct," "omission error," and "commission error" values, a task that in many cases might best be left to the discretion of the ultimate user of the information.

To avoid these problems and still collect meaningful data, it was decided to require the interpreter merely to grid agricultural areas as shown on the photography into square-mile cells (thus making possible direct comparisons with ground data on the 32 sample plots) and to tabulate estimates of the acreage of barley and wheat in each cell without regard to the specific location of individual fields.

The agricultural area within Maricopa County was divided into three nearly equal portions, and one interpreter was assigned to each area. The interpreters, chosen on the basis of high performance on preliminary photointerpretation tests, were trained using photos and ground data maps of areas they would not interpret later. Training included both identification of wheat and barley and estimation of field acreage. The interpreters were then supplied with (1) Ektachrome photos of their test areas taken on May 21 and June 16, 1970 (scale 1/120 000), and (2) maps indicating township boundaries. Each township (nominally a 6-mi square), was located on the test photography and interpreted as a unit, section by section. For each section (1 sq mi) the interpreter recorded total acreage of wheat, barley, and all cropland. (Deductions from cropland were made for farmhouse-barn complexes, freeways, major canals, and general urban and developed areas, but not for secondary service roads or local irrigation ditches.) In addition, each interpreter was asked to interpret one township in another interpreter's area and to repeat the interpretation of one township in his own area without reference to his earlier results.

RESULTS

The crop identification and acreage estimation results for each interpreter were compiled as follows:

(1) Each interpreter's estimate of acreage for the categories of barley, wheat, wheat and barley combined, and total cropland within the sample plots in his area compared with the actual acreages for each of the plots as determined by ground surveys.

(2) Ratios of actual acreages to interpretation acreages for each category were calculated for each interpreter, and this ratio was used to adjust the results for the *entire* area as estimated by each interpreter by the equation

$$\hat{Y}_I = Y_{PI} \times R$$

where

$\hat{Y}_I$ = estimate of total acreage of category within an interpreter's area

Y_{PI} = initial photointerpretation of acreage within an interpreter's area

R = the correction ratio as derived from the sample plots

(3) The category estimates for the three interpreters were summed to form a total county estimate.

(4) Sampling errors were calculated for the various category estimates by each interpreter as well as for the overall county estimates in order to give an indication of the accuracy of the crop estimates. In calculating the overall county statistics, each of the three interpreters' areas was handled as an individual stratum.

A summary of the survey results is presented below (tables 9–1 through 9–4). The sampling error is presented as a percentage figure calculated by the following equations

$$\text{sampling error percent} = \frac{S_{\hat{Y}}}{\hat{Y}}$$

where

$S_{\hat{Y}}$ = standard error of the estimated acreage

$\hat{Y}$ = estimated acreage

TABLE 9–1.—*Acreage Estimates and Sampling Error*

Category	Total estimate, acres	Sampling error, percent
Barley	50 044	11
Wheat	41 714	13
Barley and wheat	92 207	8
All cropland	452 000	3

TABLE 9–2.—*Ratio Correction Factors*

Interpreter	Barley	Wheat	Barley and wheat	All cropland
1	1.1225	0.9846	1.0481	0.9913
2	1.1131	.9012	1.0352	.9809
3	1.1234	.9388	1.0309	1.0094

TABLE 9–3.—*Sampling Error of Interpreters*

Interpreter	Error per crop, percent			
	Barley	Wheat	Barley and wheat	All cropland
1	18	17	14	5
2	30	32	16	3
3	14	21	11	6
Total area	11	13	8	3

TABLE 9–4.—*Interpretation Time*

Interpreter	Training time		Interpretation time		Average time per township	
	hr	min	hr	min	hr	min
1	8	55	26	20	1	20
2	7	30	13	40	1	03
3	6	30	28	05	1	02
Total	22	55	68	05	1	08

A correction ratio greater than 1 indicates that the interpreter underestimated the acreage of that category, whereas a ratio less than 1 indicates that he overestimated the acreage.

The results of greatest interest are the accuracies of the estimated acreages for each category in the county. However, there are no reliable statistics gathered in the conventional manner with which to compare these results. Although the Statistical Reporting Service does publish monthly estimates of crop acreages for the United States as a whole and for individual States, their methods are such that no accurate estimates are available for specific counties until months after the harvest, and even then they are much less accurate than the State and national estimates. This serves to emphasize the potential value of estimates obtained by means of photointerpretation and estimation. It is possible, however, to discuss the accuracy of the estimates by reference to calculated measures of statistical reliability derived from the sample data.

The sampling error (standard error of the estimate expressed as a percent of the estimate) for barley was 11 percent and for wheat was 13 percent, whereas the figure for both barley and wheat combined was 8 percent; indicating that a good deal of error resulted from a confusion of the two grain crops. This same phenomenon is evident in the correction ratio figures. In general, the interpreters underestimated the acreage planted to barley and overestimated wheat, but they were only slightly low in their estimates of the two grains combined. These results indicate that considerable improvement in the acreage estimates could be realized if a more definite differentiation could be made between the two grains. Nevertheless, the accuracies of acreage estimates as shown are quite encouraging, especially considering the rapidity with which the data were produced, the relatively large area interpreted, and the lack of any other reliable estimates with which they could be compared.

Table 9–3, which lists the interpreters' accuracy levels, shows that one of the interpreters had a significantly higher error for both barley and wheat than the errors of the other two interpreters, but all three were nearly equal for barley and wheat combined. This indicates that although this one interpreter had more trouble differentiating between

the two crops, he did nearly as well as the others in distinguishing the two small grains from all other field conditions. Furthermore, the large differences in performance point up the importance of screening and training interpreters before undertaking operational surveys. The sampling error could have been significantly reduced if the performance of the less accurate interpreter had been comparable to the other two. All three interpreters indicated that their confidence in their interpretations increased as they progressed through the survey. Certainly any fully operational survey would include considerably more interpreter training than was undertaken in this study.

CONCLUSION

The stated purpose of the experiment was to investigate the feasibility of performing semioperational inventories of agricultural resources using very small scale aerial or space photography. Further, it was hoped that if experimenters were cognizant at all times of the constraints that would be faced when carrying out an operational survey, findings would be more valuable than those from the more usual limited-area tests.

The results to date are encouraging on two counts: (1) the very practical problems of an operational survey are being faced and solutions are being found, and (2) it would seem that a fully operational agricultural inventory using space photography is not beyond the scope of present technology.

The biggest problems that will be faced in establishing a functional inventory system are those concerning logistics and data handling. For example, it will be necessary to insure that ground crews are at the proper place at the proper time over widely scattered areas to provide calibration data. Imagery must be obtained at specific times to permit differentiation among various crop types; interpretation of large areas must be performed rapidly to insure that the information is not outdated before it is available; and interpretation results must be compared with calibration data and the necessary adjustments made before distribution.

Finally, crop statistics must be provided, not necessarily at those times and for those geographic units that lend themselves well to the data-gathering techniques, but rather at times and for area units that are geared to user requirements as nearly as possible.

Most of the data-handling problems associated with photo inventories are not much more complex than those faced by Government agencies gathering agricultural data by conventional means at the present time. Furthermore, a number of systems are being developed, which, it is hoped, will automatically extract information from aerial or space photographs, perform crop identification functions, combine this information with other parameters keyed to the same geographic coordinate system, and produce graphical or tabular output in a wide variety of desired formats. It appears that such systems would lend themselves particularly well to agricultural surveys wherein nearly all the image interpretation is based on tone or color discrimination (a function much more accurately performed by a machine than a human interpreter) rather than complex deductive decisions. In fact, it is planned that further studies of agricultural inventory method by the Forestry Remote Sensing Laboratory will involve an investigation of the extent to which automatic image interpretation and data-handling methods can contribute to operational surveys of the type described in this paper.

SELECTED LITERATURE

CARNEGGIE, D. M.; L. R. PETTINGER; C. M. HAY; AND S. J. DAUS. 1969. Analysis of Earth Resources in the Phoenix, Arizona, Area. *In* R. N. Colwell, et al., An Evaluation of Earth Resources Using Apollo 9 Photography. Final Rept. NASA Contract no. NAS–9–9348. Univ. of California, Berkeley.

JOHNSON, V. W., ET AL. 1969. A System of Regional Agricultural Land Use Mapping Tested Against Small Scale Apollo 9 Color Infrared Photography of the Imperial Valley. U.S. Dept. Int. Status Rept. III, Tech. Rept. V, Contract no. 14–08–0001–10674. Univ. of California, Riverside.

PETTINGER, L. R., ET AL. 1969. Analysis of Earth Resources on Sequential High Altitude Multiband Photography. Spec. Rept. Forestry Remote Sensing Laboratory, Univ. of California, Berkeley.

U.S. DEPT. COMM., URBAN RENEWAL ADMINISTRATION. 1965. Standard Land Use Coding Manual.

10

Significance of the Results Obtained in Relation to User Requirements

ROBERT N. COLWELL

The preceding chapters have indicated that (1) a few important Earth resource features are consistently identifiable, even on individual frames of black-and-white photography taken by the Apollo 9 astronauts; (2) many more Earth resource features are identifiable on the matching frames of Infrared Ektachrome photography; and (3) the accuracy with which these features can be identified is improved in some instances through the use of optical or electronic equipment which produces a single color-enhanced image from two or three matching frames of multiband black-and-white photography.

In addition, the preceding chapters indicate that there must be some exciting possibilities for acquiring still more information about Earth resources through the use of "sequential" space photography. This latter conclusion must be largely inferential, of course, because all photography of our test sites that is sequential to the Apollo 9 mission has been from aircraft rather than spacecraft. Nevertheless there is strong evidence for this claim as to the value of sequential space photography, especially in chapters 3 and 4.

Encouraging though these results may be, the present report would be grossly incomplete if no attempt were made in it to answer the obvious question, "So what?" More specifically, three questions need answering:

(1) Is there really a significant user requirement for each kind of information shown in this report to be derivable from multiband aerial and space photography?

(2) What other inventory data might the user need beyond that which this report indicates might be obtainable directly by multiband remote sensing from aircraft or spacecraft?

(3) Might acquisition of the other inventory data be facilitated through certain uses of multiband aerial and space photography beyond those already described in this report?

Although the present chapter does not give a complete answer to these questions, it seeks to answer them in sufficient detail to give the balanced presentation which otherwise this report most certainly would lack.

APPROACH TO THE PROBLEM

In seeking to relate any remote-sensing capabilities (such as those of the Apollo 9 S065) to user requirements, etiher of two approaches might be used.

In the first approach, remote-sensing capabilities would be considered at the outset and, in the light of these capabilities, an exhaustive list would be compiled showing all the kinds of data that might be attained through the full exercise of these capabilities. Then due consideration would be given to each item on the list to determine whether that item might conceivably satisfy some user requirement.

In the second approach, economically significant or otherwise important user requirements would be listed. In compiling the list, the investigator would take pains to determine the informational requirements of all the agencies and types of individuals who might conceivably be served. Once the list had been compiled, consideration would be given to the various remote-sensing capabilities in an effort to determine which of these requirements might be met and by what remote-sensing process.

If either of these two approaches were used, however, consideration would eventually need to

152

be given to a compromise between user requirements and remote-sensing capabilities. Thus, for example, under the second approach, if one type of information which had been specified as needed could not be attained by remote sensing, the investigator would consider whether some modification of the information requirement, which he would be willing to accept, would indeed be derivable through the remote-sensing process.

The treatment appearing later in this chapter follows the second of these two approaches in that user requirements are first listed and remote-sensing capabilities are then considered. Emphasis is placed primarily, but not exclusively, on the vegetation resource. Without some such limitation, the analysis would initially be so complex as to be unmanageable within the framework of this report.

Even before entering into that analysis, however, it is important to acknowledge that other workers recently have directed their attention to one aspect or another of the problem of maximizing the usefulness of space photography for Earth resource inventories. One such paper, entitled "Let Aircraft Make Earth Resource Surveys" (Amrom Katz, 1969), had a subtitle that read as follows: "What we need from Earth-resource surveys can be gained easier, better, sooner and cheaper using aircraft rather than spacecraft, and in a politically more palatable and manageable manner." At the same time that this all-encompassing statement was being made, other persons were enthusiastically proclaiming that the Apollo 9 photographic experiment, with which this present report deals, showed that resource surveys of the future could and should be made from spacecraft rather than aircraft.

From the outset, those of us who have been working on the NASA Earth Resources Survey program have considered it probable that neither of these alternatives would provide the optimum solution. Consequently, much of our research has been designed toward developing a "multistage" sampling technique whereby the Earth resource inventory would be performed using three data-collecting systems: satellites, aircraft, and ground observers, in that sequence. Each of these in turn would provide progressively closer looks at progressively smaller areas, and would provide progressively more detailed information about those areas. Then, the more detailed information would,

in each instance, be applied to a much larger area for which the limited sample appeared to be representative, as evidenced by the similarity of that area to certain surrounding areas, as seen on aerial and space photographs. An excellent test of this multistage sampling technique, indicating the great promise which it holds for the inventory of Earth resources, has been performed by Philip G. Langley of the Pacific Southwest Forest and Range Experiment Station of the U.S. Forest Service, as described in chapter 8 of this report. In that study it was determined that, for an area of 6 million acres in the Mississippi Valley, a 58-percent reduction in the sampling error was attributable to information obtained from the Apollo 9 photography.

Similar multistage sampling studies, based on Apollo 9 photography, are being made by other members of the NASA-financed forestry team, some of whom are based at Corvallis, Oreg., and others at Fort Collins, Colo. The report by C. E. Poulton, R. S. Driscoll, and B. J. Schrumpf (1969) illustrates a multistage sampling technique whereby more range resource information is obtained at less cost by first making broad rangeland classifications on space photography and then making more detailed classifications through the use of aerial photography and ground observations.

USER REQUIREMENTS

As we begin a consideration of user requirements for information about the vegetation resource, table 10–1 provides a useful point of departure.

Judging from that table there are only four main categories of vegetation for which information is sought, viz, agricultural crops, timber stands, rangeland vegetation, and brushland vegetation. The left-hand column of table 10–1 shows that, by and large, the users of agricultural crop data need only six categories of information; i.e., crop type, crop vigor, crop-damaging agents, crop yield per acre by type, crop acreage by type, and total yield. The other three columns of table 10–1 show that essentially these same six categories of information are likewise the ones sought by the managers of timberlands, rangelands, and brushlands.

Table 10–2 gives the many different agencies and groups desiring information about the vegetation resource. We can logically list these many users under the same four column headings as were used in the previous table, viz, the agencies

TABLE 10-1.—*User Requirements for Vegetation Resource Data: Type of Information Desired*

Agricultural crops	Timber stands	Rangeland forage	Brushland vegetation (mainly shrubs)
Crop type (species and variety)	Timber type (species composition)	Forage type (species composition)	Vegetation type (species composition)
Present crop vigor and state of maturity	Present tree and stand vigor by species and size class	Present range readiness (for grazing by domestic or wild animals)	Vegetation density
Prevalence of crop-damaging agents by type	Prevalence of tree-damaging agents by type	Prevalence of forage-damaging agents (weeds, rodents, diseases, etc.) by type	Other types of information desired will depend upon primary importance of the vegetation (whether for watershed protection, game habitat, esthetics, etc.)
Prediction of time of maturity and eventual crop yield per acre by crop type and vigor class	Present volume and prediction of probable future volume per acre by species and size class in each stand	Present animal-carrying capacity and probable future capacity per acre by species and range condition class in each forage type	
Total acreage within each crop type and vigor class	Total acreage within each stand type and vigor class	Total acreage within each forage type and condition class	Same as above
Total present yield by crop type	Total present and probable future yield by species and size class	Total present and probable future animal-carrying capacity	Same as above

154

TABLE 10-2.—*User Requirements for Vegetation Resource Data: Agencies and Groups Desiring the Information*

Agency or group	Agricultural crops	Timber stands	Rangeland forage	Brushland vegetation
Federal agencies	Agricultural Stabilization and Conservation Service; Cropland Conservation Program; Conservation Reserve Program; Agricultural Conservation Program; Emergency Conservation Measures Program; Commodity Credit Corp.; Agricultural Marketing Service; Statistical Reporting Service; Economic Research Service; Soil Conservation Service; Federal Crop Insurance Corp.; Farmers Home Administration; Rural Community Development Service; Foreign Agricultural Service; Famine Relief Program; Foreign Economic Assistance Program; Dept. of Commerce Agricultural Census Program	U.S. Forest Service; Bureau of Land Management; plus many Federal agencies listed in col. 2	U.S. Forest Service; Bureau of Land Management; plus many Federal agencies listed in col. 2	Primarily U.S. Forest Service and Bureau of Land Management
State and county agencies	Agricultural Extension Service; State Tax Authority	Division of Forestry; Forest Extension Service; State Tax Authority	Livestock Reporting Service; Range Extension Service; State Tax Authority	Division of Forestry; Division of Beaches and Parks; Water Resource Agency; State Tax Authority
Private groups	Producers of fertilizers and pesticides; crop harvesting industry; food processing and packing industry; transportation industry; food and fiber advertising and marketing industry	Producers of fertilizers and pesticides; logging industry; wood processing industry; transportation industry; wood and wood products advertising and marketing industry	Producers of fertilizers and pesticides; meatpacking industry; tanning industry; transportation industry	Hunting and fishing clubs; public utilities commissions; local irrigation districts

and groups concerned primarily with agricultural crops, timber stands, rangeland forage, and brushland vegetation.

The array of users listed in table 10–2 is truly a formidable one; yet that list is by no means a figment of the imagination. To the contrary, each group or agency listed there presently uses a great deal of information about vegetation resources, although at present most of the information is *not* obtained from either aerial or space photography. These facts are documented in about 10 000 thoughtfully chosen words in the 1966 report by Sattinger and Polcyn entitled "Peaceful Uses of Earth Observation Spacecraft."

The complexity of satisfying requirements for the many users is not so much attributable to the formidable length of the list appearing in table 10–2 as it is to the fact that these many users want the information for different vegetation groupings, at different times, and with differing levels of accuracy. In addition, they have differing requirements as to the speed with which vegetation information must be processed once the raw data have been collected, and also as to the frequency with which the information must be updated.

This latter consideration has led to our compiling a third table (table 10–3) in our effort to document in concise, tabular form the various user requirements for vegetation resource data.

In this table (as in tables 10–1 and 10–2), the same headings can be used for the four vertical columns. At the risk of some oversimplification, this table lists six time intervals that are indicative of the frequency with which various kinds of information about the vegetation resource are needed (10 to 20 min; 10 to 20 hr; 10 to 20 days; 10 to 20 months; 10 to 20 yr, and 20 to 100 yr).

In considering relationships between the frequency with which Earth resource data should be collected and the rapidity with which these data should be processed, the writer has found it useful to employ the term "half-life" in much the same way it has been employed by radiologists and atomic physicists. The shorter the isotope's half-life, the more quickly a scientist must work with it once a supply of the isotope has been issued to him. One half-life after he has acquired the material, only half of the original amount is still useful; two half-lifes after acquisition only one-quarter of the original amount is still useful, etc.

By coincidence or otherwise, this half-life concept applied remarkably well to nearly every item listed in table 10–3. Specifically, if the frequency with which any given item of information, as listed in that table, is divided by 2, a figure is obtained indicating the time after data acquisition by which that particular item of information should have been extracted from the data. It is true that some value will accrue even if that item of information does not become known until somewhat later. But the rate at which the value of the information "decays" is in remarkably close conformity to the "half-life" concept.

Each item that is listed in table 10–3 can be placed in one of the six categories of information listed in table 10–1 and identified as a user requirement for one or more of the agencies listed in table 10–2. Thus a certain unity can be found in these three tables. Consequently, by studying them, not just individually but in concert, we can better appreciate the true nature of the multifaceted user requirement that we seek to satisfy by the remote sensing of vegetation resources.

As previously indicated, there is a high degree of complexity in the user requirement for information on the vegetation resource. Consequently, the three tables would probably be considered by many of these users to be a gross oversimplification. For example, many users think almost entirely in terms of protecting the vegetation resource from damaging agents and thus would view the problem somewhat more narrowly than we have viewed it here. On the other hand, there are those who think of the vegetation as only one of many items that comprise the total "resource complex" in a given land area they must manage. Almost certainly they would view the problem more broadly than we have in these tables, pointing out the importance of such nonvegetational components as soils, water, minerals, wildlife, and recreational potential.

An example of each of these two viewpoints will now be given, the first pertaining to agricultural resources and the second to forest resources.

Agriculturists recently were asked to list, crop by crop, the specific applications of remote sensing that might prove profitable in terms of cost/benefit ratios and total savings that might be effected. This resulted not only in their listing the six categories of information appearing in table 10–1 but also in their selecting the most important

TABLE 10–3.—*User Requirements for Vegetation Resource Data: Frequency With Which the Information Is Needed*

Time interval	Agricultural crops	Timber stands	Rangeland forage	Other vegetation (mainly shrubs)
10 to 20 min	Observe the advancing waterline in croplands during disastrous floods. Observe the start of locust flights in agricultural areas	Detect the start of forest fires during periods when there is a high fire danger rating	Detect the start of rangeland fires during periods when there is a high fire danger rating	Detect the start of brushfield fires during periods when there is a high fire danger rating
10 to 20 hr	Map perimeter of ongoing floods and locust flight. Monitor the wheat belt for outbreaks of blackstem rust due to spore showers	Map perimeter of ongoing forest fires	Map perimeter of ongoing rangeland fires	Map perimeter of ongoing brushfield fires
10 to 20 days	Map progress of crops as an aid to crop identification using crop calendars and to estimate date to begin harvesting operations	Detect start of insect outbreaks in timber stands	Update information on range readiness for grazing	Update information on times of flowering and pollen production in relation to the bee industry and to hay-fever problems
10 to 20 months	Facilitate annual inspection of crop rotation and of compliance with Federal requirements for benefit payments	Facilitate annual inspection of firebreaks	Facilitate annual inspection of firebreaks	Facilitate inspection of firebreaks
10 to 20 yr	Observe growth and mortality rates in orchards	Observe growth and mortality rates in timber stands	Observe signs of range deterioration and study the spread of noxious weeds	Observe changes in edge effect of brushfields that affect suitability as wildlife habitat
20 to 100 yr	Observe shifting cultivation patterns	Observe plant succession trends in the forest	Observe plant succession trends on rangelands	Observe plant succession trends in brushfields

NOTE.—Examples only are given. To convert this table to rapidity with which information is needed, use "half-life" concept. (See text.)

"candidate problems" in U.S. agriculture—all dealing with damage that is done each year to specific crops by specific insects or pathogens. These experts agreed that, if remote sensing would permit early detection of attacks by these insects or pathogens, effective control measures could be taken, thereby avoiding great economic loss. The most important items on this long list of agricultural candidate problems are as follows: (1) control of leaf and stem rust on wheat and oats, (2) barberry bush eradication, (3) cereal leaf beetle eradication, (4) imported fire ant eradication, (5) soybean cyst nematode control, (6) control of burrowing nematode disease on citrus, (7) phony peach and peach mosaic eradication, (8) control of various root-rot conditions on agricultural crops and timber stands, (9) control of gypsy moth infestations on pines, (10) control of wooly aphid on conifers, (11) control of pine bark beetles, (12) control of spruce budworm infestation, (13) control of conifer limb and stem rust, (14) control of hardwood dieback, and (15) control of dwarf mistletoe on conifers.

The previously mentioned complexity results from the fact that each of these candidate problems poses its own remote-sensing specifications in terms of sensors to be used, geographic areas to be sensed, optimum dates for sensing, signatures to be recognized, rapidity with which data must be analyzed, and format in which data should be presented. Obviously a definitive study needs to be made soon to determine the most suitable interaction between remote-sensing personnel and these various users of remote-sensing data. Only by this means can we properly match remote-sensing capabilities to user requirements in agriculture.

Foresters also have decided that they need essentially the same six categories of information as the agriculturists require, as seen in table 10-1. Then, using a somewhat different approach, they have decided that remote sensing might be especially helpful to them by providing information on which to base "multiple use" decisions relative to each part of the forest. The multiple-use concept is a complex one, and it is far more applicable in forestry than in agriculture. Some users of the forest would like to have all parts of the area managed primarily with a view to maximizing timber production. At the other extreme are those who want to preserve our forests as primeval museums to be

enjoyed by posterity. In between are those who would condone each of these uses for specific parts of the forest so long as it did not interfere with their objectives of using the forest primarily as a source of water for domestic use; or of minerals for industrial use; or of fish, game, boating, hiking, skiing, etc., for recreational use. The true complexity of the forester's decisionmaking problems as he seeks to satisfy these many user requirements, is suggested merely from a reading of the forester's official definition of the term "multiple use" (Multiple Use Act, 1960).

> The management of all the various renewable resources of the national forests so that they are utilized in the combination that will best meet the needs of the American people; making the most judicious use of the land for some or all of these resources or related services over areas large enough to provide sufficient latitude for periodic adjustments in use to conform to changing needs and conditions; that some land will be used for less than all the resources, and harmonious and coordinated management of the various resources, each with the other, without impairment of the productivity of the land, with consideration being given to the relative values of the various resources, and not necessarily the combination of uses that will give the greatest dollar return or the greatest unit output.

This definition is so complex that the concept of multiple use may easily escape the average reader. However, Draeger (1970) has rendered a service by providing the following more comprehensible statement of this concept:

> The allocation of lands to various uses such that the combination of uses best meets the needs of those for whom the land is being managed, provided that in the process of developing this allocation, values and costs of each of the many possible patterns of exclusive uses and compatible co-uses are at least taken into consideration. In some instances, depending on the circumstances, several uses may be derived simultaneously from the same parcel of land, while in other cases a particular area may be allocated to only a single use but managed as a portion of a greater whole composed of many such exclusive-use parcels.

In terms of user requirements for information, each of these statements implied that before an intelligent decision can be made regarding how best to use each part of the forest, information of two major types is needed:

(1) An accurate "in-place" delineation of both

the forest vegetation and all associated resources (soils, water, minerals, topography, etc.); i.e., a complete "resource map"

(2) Adequate sociological and technological data to know how the forest as a whole, and each unit of it in particular, can be made to produce "the greatest good for the greatest number" (to use the words of Gifford Pinchot)

Of these two complex and interrelated problems, the present report deals only with the first.

As we conclude this discussion of user requirements for information about vegetation resources, three summarizing points seem worthy of emphasis:

(1) Tables 10–1 through 10–3 set forth most of these requirements, in terms of types of information desired, agencies and groups requiring the information, frequency with which the information is desired, and rapidity with which the data should be processed.

(2) There are certain instances, as exemplified by the 15 candidate problems of agriculturists that we have listed, when we will be told that our three tables may have set forth the user's requirements too broadly, since interest centers in only a few vegetation resource problems.

(3) There are other instances, as exemplified by the multiple-use problems of foresters, when we will be told that our three tables may have set forth the user requirements too narrowly, since multiple-use decisions depend not merely on vegetation resources but on the entire complex of Earth resources (vegetation, soil, water, minerals, topography, etc.).

REMOTE-SENSING CAPABILITIES BASED ON AN ANALYSIS OF THE APOLLO 9 AND SEQUENTIAL HIGH-ALTITUDE PHOTOGRAPHY

The purpose of this section is to indicate the extent to which an analysis of either multiband high-altitude photography or multiband space photography of the type obtained on the Apollo 9 mission will enable photointerpreters to satisfy the informational needs of various users.

Of the three tables appearing in the preceding section, we note that table 10–1 gives the types of information desired about vegetation resources; table 10–2, the agencies and groups desiring that information; and table 10–3, how often and how rapidly they need it.

In the interest of simplifying the analysis in the present section, we will deal with each of these tables separately.

Because table 10–1 deals with the types of information desired, it is especially appropriate that we first consider remote-sensing capabilities in relation to that table. Until that has been done little is accomplished by entering into detailed discussions as to who wants the information or how soon and how frequently they need it.

Each of the six categories of information listed in table 10–1 as being desired by those who manage agricultural crops, timber stands, and forage and brush resources, will now be discussed in turn. In each instance needs of the agriculturist will be considered first, after which similar needs of those who manage timberlands, rangelands, and brushlands, respectively, will be considered.

Species Identification

Agricultural Crop Species

Chapters 3 and 4 of this report deal to a very large extent with possibilities for identifying crop types (species) on multiband space photography. Before analyzing those results we must recognize that it would be a poor farmer, indeed, who did not know what type of crop he was growing in each field that he was cultivating. However, this has proved to be of little help to the various Federal agencies and other regionally or nationally oriented groups that need to know quickly, accurately, and perhaps even discreetly, the acreages within vast areas that have been planted to various kinds of crops in a given year. Returns from questionnaires that these agencies send to farmers can be of value; but for compiling broad regional statistics, these returns generally prove to be too meager, too late, and on many occasions too inaccurate to satisfy agency needs. Consequently, each of these agencies has long been interested in obtaining such information from aerial photography and, more recently, from space photography.

In this regard, our initial look at the results reported in chapters 3 and 4 can be quite discouraging. Judging from tables appearing in those two chapters, the overall accuracy with which crop types can be identified on multiband aerial or space photography of the Phoenix and Imperial

Valley areas ordinarily does not exceed 60 percent. While accuracies greater than 90 percent may be achieved for some crops (by using the most favorable film-filter combinations), these seem to be offset by instances in which accuracies are as low as 30 percent. Far more encouraging results are indicated, however, when the same raw data as were used in obtaining these figures are examined more closely, bringing to light the following factors:

(1) Inexperienced photointerpreters tend to make the most errors and thereby lower the overall average for accuracy of crop identification. To illustrate both the importance of this factor and at least a potential remedy to the problem, let us presume that we were to know, say in the year 1975, that 45 photointerpreters would be needed that year, periodically throughout the growing season, to study space photography and identify agricultural crops throughout the United States. Instead of training the first 45 people who happened to be available for the task, we might be more selective of trainees and at the end of the training session retain only the top performers for use in the operational surveys. If by this process we found it necessary to train 150 candidates to have a sufficiently large population from which to select 45 top performers, this still might be the most efficient and economical procedure in the long run.

(2) Even among the top performers, accuracy should improve with time on the job. On similar classification projects (but using aerial rather than space photos), the U.S. Forest Service has found that trainees require nearly 6 months of intensive on-the-job training before they are able to achieve maximum performance.

(3) When individual crop types cannot be identified on the photography, the mere identification of groups of crops may provide an adequate solution. The validity of this assertion is best documented by citing results of an entirely separate research project that recently was conducted and which employed high-flight photography. A major objective of that project was to determine total wheat acreage in each of several very large areas. Those of us working on the project soon found that wheat could not be consistently identified on this photography because of the frequency with which it was confused with other "small grains" (oats, barley, and rye) grown in the area. However, it was established (from an examination of records obtained by ground enumerators in each of several previous years) that the ratio of "wheat acreage" to "total small-grain acreage" tended to be remarkably constant, year after year, in each of these areas. Consequently, the photointerpreters merely determined total small-grain acreage and multiplied this figure by the proportionate wheat acreage factor to determine total wheat acreage. Subsequent ground-truth data showed that the average error in estimating wheat acreages by districts was less than 5 percent and the maximum error was less than 10 percent.

(4) It is probable that far more image enhancements could be profitably made from the high-altitude and Apollo 9 multiband photos than have been made to date. Most of the optical and electronic enhancements that were tested in chapters 3 and 4 sought to enhance overall interpretability, thereby facilitating the identification of several different types of crops with merely one type of enhancement. Judging from results our group previously has obtained when working with multiband aerial photography, a series of enhancements can be made quickly and easily, each of which, in turn, serves to differentiate some particular crop type from everything else. Thus, enhancement 1 may give a unique coloration for crop A and thus permit it to be distinguished readily from everything else; then, merely by rotating the filter wheels of an optical combiner (fig. 2-1) or the electronic control knobs of an image discrimination, enhancement, combination, and sampling (IDECS) or Philco-Ford device (figs. 2-2 and 2-3), enhancement 2 can be produced to give a unique coloration for crop B versus everything else; a third position gives enhancement 3, which facilitates differentiating crop C from everything else, etc.

(5) On operational surveys, few areas would exhibit the variety of crops and hence the classification problems encountered in the Phoenix and Imperial Valley areas. Some of this Nation's largest crop-producing regions are devoted almost entirely to the production of two or three major crops (rice and sugarcane in one region, cotton and tobacco in a second region, corn and soybeans in a third region, etc.). Usually it is far easier for the photointerpreter to differentiate a particular crop from two or three other associated crops than from eight to 10.

(6) On operational surveys of agricultural areas, a month more favorable than March (when Apollo 9 flew) could be selected as the optimum time for photo acquisition. From a reading of chapters 3 and 4, it is apparent that three of the most prevalent crops in the Phoenix and Imperial Valley areas (sugar beets, barley, and alfalfa) have very similar appearances on small-scale photography during the month of March. It also is apparent from those two chapters that these three crops are far more identifiable later in the season. Numerous examples of this fact are provided by table 3–16 in chapter 3, for example. Both test 7 as listed in that table and test 9 were conducted on Infrared Ektachrome high-altitude photographs taken in March and May, respectively. Whereas in March the three main crops—barley, mature alfalfa, and sugar beets—could only be identified to accuracies of 33, 56, and 46 percent, respectively, in May the corresponding accuracies were 90, 83, and 40 percent.

Timber Species

In chapter 7 several Apollo 9 photographs from the Vicksburg, Miss., area are discussed. It is obvious from a study of those photographs that a photointerpreter would experience little difficulty in differentiating between forested areas and other areas on space photography of this quality.

On the Infrared Ektachrome space photography it also is possible to differentiate between the blue-gray coloration of bottomland hardwood stands and the purple-to-pink coloration of pine-hardwood stands. As documented in chapter 8, even this limited stratification of the timber resource greatly improves sampling efficiency. Furthermore, it is probable that, by applying electronic image enhancements or microdensitometer traces to the multiband Apollo 9 photographs, still further stratification of timber types would be possible.

Although not included in the present report, Apollo 9 multiband photographs also were obtained of sizable forested areas east of the Mississippi River, including those in the vicinity of Atlanta, Ga. Useful stratifications of the timber resource were possible in those areas also, but the photointerpreters experienced greater difficulty there than in the Vicksburg area.

Just as sequential photography aids in the identification of agricultural crops, such photography also should facilitate the identification of timber species. However, to exploit that possibility, the photointerpreter would need to develop a calendar of seasonal phenology for each tree species, analogous to the crop calendars presented in chapters 3 and 4 of the present report.

An additional possibility for mapping timber types on space photography is presented in chapter 6, which deals with an area having high topographic relief. In that area, south-facing slopes consistently provide a site where oaks can reach their highest elevational limits. North-facing slopes are protected from the Sun; the snow remains longest in these areas with the result that they support pines and firs. It is on these slopes that conifers are able to occupy their lowest elevational positions. Similar vegetation-soil-aspect relationships hold very consistently on all areas of Arizona and New Mexico covered by the Apollo 9 photos.

Rangeland and Brushland Species

Virtually all rangelands and brushlands suitable for study on the Apollo 9 photographs are in arid desert lands of Arizona and New Mexico. A few alpine meadows that have excellent forage during the summer months also were photographed, but at the time of the Apollo 9 overflights these areas were both snow covered and largely devoid of forage, thereby precluding the possibility of making a study of them.

Even in the arid desert lands, herbaceous forage was so sparse and brush so nearly leafless when the Apollo 9 photos were taken that it registered very poorly, if at all, on the space photography. However, in the very situation where vegetation is least conspicuous, landforms are most conspicuous. Consequently, an attempt was made, especially in the rangeland and brushland areas southwest of Phoenix, to exploit the known correlation that exists between landform type and vegetation type. Results of that effort, although thus far somewhat superficial, have been highly encouraging, as documented in chapter 3 (fig. 3–23, Color Plate 10, and in the supporting text). A similarly successful correlation was established in the vast desert area southeast of Tucson, where, as reported in chapter 5, the large shrubs and trees were confined primarily to large drainage channels, and the most xerophytic species to the interfluves, while species having intermediate drought resistance occupied intermediate environments between these two extremes.

Changes in phenology with season often promise possibilities for the photoidentification of rangeland and brushland species just as they do for agricultural and timber species. For example, chapter 5 documents the following facts as applied to both the Sonoran Desert and the Chihuahuan Desert regimes:

(1) During the second week in March, when the Apollo 9 photographs were taken, only the coniferous forests and the evergreen species in the chaparral, oak, and juniper-oak woodlands have green foliage and, therefore, register red on Infrared Ektachrome space photography.

(2) Species in the desert-shrub area are leafed out in April. Hence, any additional area that appears red on April photography usually will be of this type.

(3) Grasslands leaf out in August, following the summer rainy season. Hence, any additional area that appears red on August photography usually will be of this type.

Determining the Vigor of Vegetation and the Identity of Plant-Damaging Agents

Agricultural Crops

In chapters 3 and 4 reference is made to the variability in crop appearance attributable to variability in vegetation density. Assuming that the crop can be identified despite this variability, a powerful tool is provided for determining both crop vigor and identity of the damaging agent through density estimation.

If the objective is merely to detect areas in which there is sparse vegetation (attributable to a lack of crop vigor), space photography taken on only a single date and with only one of several film-filter combinations probably will suffice. But if, in addition, a determination of the damaging agent is to be made, the photointerpreter's task is greatly facilitated by his having access to sequential photography taken with a multiband camera system at two or more dates during the growing season.

Of the many examples that might be given to illustrate this fact, only one will be offered here. In the Imperial Valley, at the time of this writing, there is a serious threat to the cotton crop attributable to a damaging agent known as "pink

worm." (Strictly speaking, this damage is attributable to a combination of pink boll worm and certain leaf borers (defoliating insects).) Heavy defoliation (and a consequent reduction in vegetation density) constitutes one of the manifestations of damage to the cotton plants.

Cotton fields having a low vegetation density should be detectable even on space photography, judging from the previously cited results of chapters 3 and 4. Even so, if a cotton field were to be photographed on only one date, pink-worm damage might easily be confused with at least two other factors that can affect the density of vegetation in a field.

One of these factors, locally known as "cultivator blight," is attributable merely to poor traversing of a cultivator down the rows of cotton. Small deviations from the prescribed course will cause the cultivator tines to remove the developing cotton plants. The other factor is variable planting depth at the time the cotton seeds are mechanically sown. In shallow-sown areas seedlings emerge from the soil and start vegetating the cotton field sooner than they do in deeply sown areas.

Given aerial or space photography that was taken on only one date during the development of plants in a cotton field, the photointerpreter may find it impossible to determine which of these three factors is responsible for the sparseness of vegetation. But if he is given photography taken on each of two dates, suitably spaced, he should be able to determine whether the area of sparse vegetation is increasing (indicating that pink worms are the damaging agent), remaining about the same size (attributable to cultivator blight), or decreasing (because of late emergence of seedlings in deeply sown areas).

Timber Crop Vigor

In several reports which our group has submitted to the National Park Service and other sponsoring agencies, we have documented the feasibility of using aerial photographs to detect loss of vigor in coniferous timber stands. In most instances the damaging agent was some species of bark beetle (e.g., *Dendroctonus brevicomis*, *Scolytus ventralis*). Several of our Forest Service colleagues, including Heller, Aldrich, Wear, Weber, and Croxton, also have reported favorably on the feasibility of detecting insect attacks in timber stands, as well as fungus attacks that produce certain economically important diseases (e.g., Douglas-fir root rot, ash dieback, oak wilt, and Dutch elm disease). Infrared-sensitive films often are best for detecting these maladies.

For the early detection of such phenomena, the photointerpreter must be able to resolve individual tree crowns, or even portions of tree crowns. Space photography such as that obtained on the Apollo 9 mission does not provide sufficiently high spatial resolution to permit early detection of vigor loss on a tree-by-tree basis.

Once a sizable portion of the timber stand has become damaged, this fact should be determinable from a study of space photography of the Apollo 9 type. However, it was not possible to document this point in our present studies for two reasons:

(1) To the best of our knowledge, such sizable infestations did not exist on any coniferous stands covered by Apollo 9 except for one area 50 mi north of Phoenix where a heavy snowfall still hung on the trees and obscured the foliage.

(2) Even if such infestations did exist on deciduous hardwood stands, they would not have been detectable because the trees were in a nearly leafless condition in March.

The initial outbreaks of forest insects and disease infestations commonly occur in timber stands that occupy specific topographic sites. Such sites usually are detectable on space photography, and once detected could be aerially photographed periodically. This is yet another example, therefore, wherein the previously described process of multistage sampling might be used to good advantage.

Forage and Brush Vigor

In the rangelands northeast of Phoenix, some of the annual grasses, herbs, and brush species were vigorously growing at the time of the Apollo 9 mission, especially at mid-elevation on the foothills. At lower elevations such vegetation tended to be growing less vigorously, mainly because of inadequate soil moisture. At higher elevations it also was growing less vigorously, mainly because cold air and soil temperatures (which characterize the winter and early spring months in this area) still prevailed. The more vigorously the vegetation is growing, the more infrared reflective it is, and therefore the redder it appears on Infrared Ektachrome space photography, as shown in Color

Plate 11. It is quite apparent, in that the mid-elevation rangelands and brushlands are the reddest. As a corollary to the above, they are at the optimum state of readiness for grazing. Like other kinds of vegetation, grasses, herbs, and shrubs can be damaged on occasion by insects and fungi. When this occurs, the same principles governing damage detection on aerial or space photographs will apply as were described in the previous section dealing with timber stands.

Yield Estimation

Agricultural Crops

To the extent that it is possible for the interpreter of space photographs to distinguish crop types, estimate crop vigor, and identify crop-damaging agents, it also is possible for him to estimate crop yields. However, it first is necessary for him to "calibrate" various photographic images of crops in terms of crop yields. When such calibrations are being developed, the dimension of time must be taken into consideration, which again points to the desirability of obtaining sequential photographs on properly selected dates. For example, in previous aerial photointerpretation studies on yield of wheat, our group found that fields that, 3 weeks before first heading, exhibit 80 percent severity of a disease known as "blackstem rust" will produce a yield that is only 10 percent of normal. However, fields that do not exhibit 80 percent severity until 1 week before first heading will produce a yield that is 90 percent of normal.

Until such calibrations have been developed, crop by crop, meaningful estimates of yield cannot be made from aerial or space photography. Our limited work in this area, using both Apollo 9 and sequential high-altitude photography, suggests that a great deal of this type of research might profitably be conducted in the near future. We are further drawn to this conclusion upon noting the emphasis given by U.S. Department of Agriculture personnel to various candidate problems (as previously listed), every one of which stems from a concern about the yield reduction imposed on some particular crop by some particular damaging agent.

Timber-Stand Yields

Foresters measure potential yields by determining volumes of merchantable timber by species in each part of the forest. Until about 1940, such information was obtained entirely by ground measurements known as "timber cruises." Then it was found that a study of aerial photos would permit the forest to be stratified into nearly homogeneous units, thereby reducing by 90 percent or more the amount of ground measurement required. Consequently, most timber-volume estimates of the past two or three decades both in the United States and in most other parts of the world have been made with the aid of aerial photos.

Photographs obtained on the Apollo 9 mission permitted foresters, for the first time, to make a realistic test of the extent to which space photography might still further facilitate the task of acquiring forest-yield data (i.e., estimates of timber volume by species). In the areas being studied, even the crude stratification of timber stands that is done on space photography can improve sampling efficiency by approximately 60 percent when the objective is to determine potential yields of timber stands throughout the forest, as indicated in chapter 8.

Forage and Browse Yields

"Animal-carrying capacity" is perhaps the most common measure of yield employed by those who manage rangelands and brushlands. Such capacity is commonly expressed in "animals per unit area per year," thereby indicating the number of adult animals of a specified type that can profitably be grazed on a specified area. As with many of the other applications discussed in this chapter, only very crude estimates of animal-carrying capacity could be made from space photography of the Apollo 9 type unless some prior "calibration" had been made on representative space photographs. The closest approach to this task that has been performed to date will be found (1) in the work of Poulton et al., part of which is described in chapter 5 of the present report, and (2) in the paper by Poulton, Driscoll, and Schrumpf (1969) entitled "Range Resources Inventory From Space and Supporting Aircraft Photography."

In most rangeland areas, whether forage is consumed in the green state or after it has dried, yield is directly proportional to the total amount of healthy vegetation that was present in the area at the peak of the growing season. In such areas, space photography may provide the means by which a very useful stratification of rangeland forage by yield categories can be made. (See, for example, the varying intensities of red coloration, indicative of variations in foliage density, that appear in the Apollo 9 photo of Color Plate 11.) This relationship may be less useful, of course, in areas where unpalatable or poisonous species comprise a variable and unknown proportion of the total vegetation.

Area Measurement

Agricultural Areas

In chapter 3 possibilities for making area measurements in agricultural areas are discussed. As indicated there, the solution of this problem is greatly facilitated if the land has been surveyed into sections (640 acres), quarter sections (160 acres), and 40-acre blocks. In that event, the net acreage in each field usually can be obtained merely by deducting from gross acreage the land occupied by roads, canals, and buildings. In areas of rolling-to-steep topography, fields may assume highly irregular shapes and may be intermingled with swamplands, brushlands, and timberlands. Numerous examples of this condition will be found in the area centered around Vicksburg, Miss., as illustrated in chapter 7. In such areas, methods described below may provide the most satisfactory solution.

Acreage Determination in Timber, Forage, and Brushland Areas

Although not discussed elsewhere in this study of Apollo 9 photography, a method commonly known as "dot apportioning" can be used in conjunction with aerial photos in areas where land-use classes can be recognized on the photos but where the exact boundaries are difficult to discern, just as on many parts of the Apollo 9 photography. That method makes use of a dot grid that has been printed on a piece of transparent cellulose acetate. The acetate is placed over the photo (either randomly or systematically, depending on the sampling design). Then, for each dot a determination of the land-use class (timber type, forage type, brush type, etc.) represented by the photo-image directly beneath the dot is made and recorded. Once an adequate sampling for area has been obtained by this means, the acreage comprising the total area is apportioned to land-use classes in accordance with dot counts.

Space photos are superior to aerial photos for making such determinations on sloping ground. Relief displacements on vertical aerial photos cause areas to appear too small if they slope away from the camera station and too large if they slope toward it. On vertical space photos, however, this problem usually can be ignored. Relief displacements are so minimal that any such photo can be treated as a map without danger of introducing significant errors in area determinations.

Determination of Total Yield

This determination is merely a mathematical problem once the space photos have provided information as previously discussed on vegetation type, crop yield, and acreage. Hence no further discussion of this item is deemed necessary in this report.

SPACE PHOTOGRAPHY CAPABILITIES IN RELATION TO THE AGENCIES

Table 10–2 lists the major U.S. agencies desiring information of the types listed in table 10–1. As explained in the report by Sattinger and Polcyn on "Peaceful Uses of Observation Spacecraft" (1966), each agency has its own preferences as to format in which the information should be presented. Most of them, however, merely are trying to get a true "picture" (to use their own words) of the resources they are seeking to manage. Alternately stated, they wish to know, in terms of the resources, "how much of what is where." Since we therefore are considering with reference to table 10–2 merely different ways in which information might be presented to each user, no remote-sensing breakthrough in this respect is required. Therefore, as we conclude our brief consideration of table 10–2 and proceed to a consideration of table 10–3, it is at least helpful to reflect on the remarkable extent to which an annotated enlargement of a space photograph of the Apollo 9 type might give each of these many users the desired resource "picture"—both literally and figuratively.

SPACE PHOTOGRAPHY CAPABILITIES IN RELATION TO THE FREQUENCY AND RAPIDITY WITH WHICH THE INFORMATION IS DESIRED

It is indicated in table 10–3 that some types of information pertaining to the vegetation resource are desired at intervals as frequent as 10 to 20 min, at least during certain critical periods and in certain critical geographic areas. Furthermore, the table indicates that for most of the benefit to be realized, such information must be extracted and disseminated to those seeking it within 5 to 10 min (i.e., within about one "half-life" after data acquisition).

While the results of our Apollo 9 test may justify optimism in satisfying some informational requirements, there are several reasons why such optimism does not appear to be justified with respect to these particular emergency or early-warning types of requirements:

(1) Since a single satellite orbiting the Earth at an altitude of a few hundred miles is within data-collecting range of a given area only for about 3 min every 3 or 4 days, a prohibitively large number of these vehicles, operating in tandem, obviously would be required to provide the frequency of surveillance desired.

(2) Presumably it is as important to be able to acquire this kind of information at night as well as in the daytime, and no nighttime photographic capability from space is currently envisaged.

(3) Even if a vehicle were over the proper area at the proper time of day, the chances are extremely poor in many areas that conditions would be sufficiently cloud free to permit reconnaissance by means of photography. (For example, one recent study indicated that during the crop-growing season, only one pass out of 28 would be cloud free for a 1000-sq-mi agricultural area in Indiana.)

(4) Automatic data-processing techniques probably would be required to process the acquired data quickly enough to permit timely delivery of the desired information to the user.

Fortunately, most of the information desired relative to vegetation resources is not needed so frequently nor so promptly as that just discussed.

Although weather will constitute a serious deterrent in many areas, we find that the farther we proceed down the lists in table 10–3, the less troublesome that factor becomes (because we have more opportunities for cloud-free weather while in orbit over the area of interest).

SUMMARY WITH REFERENCE TO THE THREE USER-REQUIREMENT TABLES

In this chapter we have stated that our analysis of space capabilities would be incomplete unless we answered three questions. We can now answer them as follows:

(1) There are, indeed, important user requirements for the various types of information shown in this report to be derivable from space photography (our discussion of tables 10–1 through 10–3 certainly has documented this fact).

(2) Certain other kinds of inventory data also are needed beyond those investigated in this report.

(3) Acquisition of much of the additional information that is desired probably can be facilitated in the future through proper use of multiband space photography of the type acquired on the Apollo 9 mission.

There still would be many items of desired information that are not attainable by this means. But it also is important to recognize that remote sensing of vegetation and other Earth resources from space still can be eminently worthwhile even if it does not solve all of the Earth resource manager's informational requirements.

SELECTED LITERATURE

CROXTON, R. J. 1966. Detection and Classification of Ash Dieback on Large Scale Color Aerial Photography. U.S. Pac. Southwest Forest Range Exp. Sta. Res. Pap. PSW–35. Berkeley, Calif. 13 pp.

DRAEGER, W. C. 1970. Applications of Remote Sensing in Multiple-Use Wildland Management. Ph.D. thesis, Univ. of California, Berkeley. 116 pp.

HELLER, R. C.; R. C. ALDRICH; W. F. McCAMBRIDGE; F. P. WEBER; AND S. L. WERT. 1968. The Use of Multispectral Sensing Techniques to Detect Ponderosa Pine Trees Under Stress From Insect or Pathogenic Organisms. U.S. Pac. Southwest Forest Range Exp. Sta. Annual Rept. Earth Resources Survey Program, Off. of Space Science and Applications, NASA.

KATZ, A. H. 1969. Let Aircraft Make Earth Resource Surveys. Astronaut. Aeronaut. 7(6): 60–68.

Multiple Use Act. 1960. PL 86–517, HR 105–72.

POULTON, C. E.; R. S. DRISCOLL; AND B. J. SCHRUMPF. 1969. Range Resource Inventory From Space and Supporting Aircraft Photography. Vol. 2: Agriculture, Forestry, and Sensor Studies. Proc. 2d Annual Earth Resource Aircraft Program Status Rev., pp. 20–1 to 20–28. NASA, MSC, Houston, Tex.

SATTINGER, I. J.; AND F. C. POLCYN. 1966. Peaceful Uses of Earth Observation Spacecraft. Vol. 2: Survey of Applications and Benefits. IROS, Willow Run Laboratories, Inst. Sci. Tech., Univ. of Michigan, Ann Arbor. 159 pp.

11

Summary and Conclusions

Robert N. Colwell

In the introductory chapter of this report, a description is given of the types of photography that were obtained (1) on the Apollo 9 mission, (2) on concurrent flights made over some of the same NASA test sites by supporting aircraft, and (3) on subsequent high-altitude flights designed to obtain simulated sequential space photographs. To set the stage for an analysis of this photography, the need for Earth resource surveys and the value of aircraft and spacecraft as the platforms from which to make such surveys are then considered. The introductory chapter also includes a presentation of the rationale for using sequential multiband photography when making Earth resource surveys.

Chapter 2 discusses and illustrates various possibilities for the enhancement of multiband photography through the use of additive-color techniques. In chapters 3 through 9, numerous aerial and space photographs obtained in connection with Apollo 9 S065 are presented and interpreted. Chapter 10 relates these results to the informational requirements of various users of Earth resource data.

Although highly encouraging results have been obtained from the studies we have made to date, it is deemed important to defer final conclusions as to operational feasibility until more quantitative tests of the type reported upon in chapters 3, 4, 7, 8, and 9 have been performed by our group and by others. In this regard, it should again be emphasized that our group is only one of several groups that are seeking to determine the usefulness of Apollo 9 photography for the inventory of various kinds of Earth resources. These other groups share our opinion, however, that the success achieved by the Apollo 9 astronauts in obtaining high-quality multiband space photography of selected NASA test sites was very timely. That photography is providing all of this Nation's Earth resource survey investigators with a much-needed opportunity to study both the opportunities that will be presented and problems that will be encountered when operational multiband satellites are launched. The multiband system scheduled for use in ERTS–A will employ almost exactly the same wavelength bands as were incorporated in the Apollo 9 multiband camera system. This fact makes even more meaningful the research currently being done on the Apollo 9 photography to establish "tone signatures" for various Earth resource features.

The aerial photographs and optical mechanical scanner records that were obtained during the Apollo 9 mission are permitting us to evaluate various multistage sampling schemes that will further enhance the usefulness of space photography for the inventory of Earth resources. The fact that sequential, very-high-altitude photography also has been obtained using the same wavelength bands and covering the same NASA test sites adds further significance to analyses currently being made of the Apollo 9 photography.

From the studies of sequential photography that we have conducted thus far, we tend to conclude that this photography will be far more useful in some Earth resource disciplines than in others.

From our studies of the Apollo 9 photography and of the supporting high-altitude photography that was obtained both at the same time and on subsequent dates at roughly 1-month intervals, we conclude the following:

(1) Although the nominal resolution of the Apollo 9 photography is only about 300 to 400 ft, based on low-contrast targets (see table 1–2), some features such as roads and canals having a least dimension of only 20 to 30 ft are clearly discernible.

(2) The nominal resolution of the supporting

168

35-mm high-altitude photography is approximately one order of magnitude better than that obtained on the Apollo 9 mission.

(3) The accurate discrimination of some important Earth resource features (e.g., timbered versus agricultural lands; vegetated versus fallow fields) is possible even on individual black-and-white space photos.

(4) A great many more Earth resource features are identifiable on Infrared Ektachrome space photos (e.g., bottomland hardwood stands, pine-hardwood stands, and certain individual crop types).

(5) The interpretability of some of these features is still further increased through the use of optical or electronic equipment to form color-composite images from two or more wavelength bands of black-and-white photography.

(6) High-altitude photography, despite its much higher spatial resolution, and even though taken with the same film-filter combinations as were used on the Apollo 9 mission, provided very little improvement in the interpretability of specific Earth resource features that were subjected to quantitative evaluation tests as described below.

(7) Detailed studies were made in several geographic areas in order to arrive at a quantitative determination of the interpretability of the Apollo 9 photographs and of the associated high-altitude photography. In these tests the photography was studied both in its original state, either as opaque prints or as positive transparencies, and after being electronically or optically enhanced by various means. One set of studies dealt with agricultural crops near Phoenix, Ariz. (ch. 3), and also in the Imperial Valley of California (ch. 4). These quantitative studies led us to the following conclusions:

(a) When the only photography made available to the photointerpreters is that taken by the Apollo 9 astronauts (in March 1969—a less-than-optimum time of year for this purpose), the accuracy with which the interpreters are able to identify crop types and field conditions rarely exceeds 60 percent. Those who wish to use this kind of agricultural information ordinarily require a much higher order of accuracy; e.g., 90 to 95 percent.

(b) If, in addition, the photointerpreters are given access to the high-altitude photography that was taken on the same date but at much higher spatial resolution, the quantitative data show that, for the particular Earth resource features being investigated, there was no statistically significant improvement in image interpretability. Given sequential high-altitude photography taken at later dates, however, the accuracy was substantially improved, as shown in the tables of chapters 3 and 4.

(c) On both the space photography and the high-altitude photography, Infrared Ektachrome was significantly more interpretable than any of the matching frames of multiband black-and-white photography.

(d) When the matching frames of black-and-white photography were combined and enhanced either optically or electronically, interpretability was improved to where the information derivable was roughly equivalent to (but only rarely better than) that obtained from the Infrared Ektachrome photography. It is to be emphasized, however, that our studies to date on this particular project have not permitted us to investigate adequately the full possibilities of multiband image enhancement. Primarily because of time limitations, the enhancements that we have studied thus far are those made with a view to improving the overall interpretability of the multiband black-and-white photography, as indeed they did. The next step will be using optical and electronic image enhancements of this multiband black-and-white photography to make an entire series of color enhancements, one of which is designed to distinguish one Earth resource feature from everything else, a second enhancement to distinguish a second Earth resource feature from everything else, etc. Some of our earlier studies, although dealing only with aerial photography, have shown that through the making of such a series of enhancements, results superior to those provided by Infrared Ektachrome photography are quite commonly obtained.

(e) In any operational system designed to exploit the broad synoptic view of space photography, it would be highly desirable to achieve high spatial resolution on the entire frame instead of merely on small sections of it, as in this report. Before this could be done, however, an optical combiner providing far better means for effecting registration than the one available to us would need to be developed; and for electronic combiners the additional need would exist for increasing the number of scan lines on the raster.

(8) Several similarities are evident from the analysis of agricultural crops in both the Phoenix and Imperial Valley areas. Because crop types and cropping patterns are much alike in the two areas, there exists a close correlation in crop calendars (figs. 3–18 and 4–9). In addition, the dates judged optimum for discrimination of specific crop types are quite close in the two areas. For example, cereal grains mature during May in both areas, and they can best be identified at that time. Cotton is planted only in fields that are bare in March (date at which Apollo 9 and high-altitude photography was first obtained for the two areas); thus, predictions of where cotton will be planted can be made by identifying bare-soil fields in March. The location of cotton can be confirmed later in the growing season.

These examples—for barley and cotton—as well as for others in chapters 3 and 4, illustrate the value of obtaining photography at more than one date during the year. Results of interpretation tests using both single date photography and multidate photography (tables 3–16, 3–17, and 4–1 to 4–5) indicate that, in both the Phoenix and Imperial Valley areas, the accuracy of identification of every crop type on high-altitude aerial photographs can be increased by studying crop sequences throughout the growing season. Similarities between the two areas also suggest that it might be possible to prepare a regional photointerpretation key for crop identification that would be applicable for both the large area of central and southern Arizona and the Imperial Valley and adjacent southern California areas that have common climatic patterns and agricultural practices.

The value of sequential analysis for wildland vegetation was also clearly demonstrated. In the Phoenix area, it is shown that judicious choice of dates for obtaining photography will enable the photointerpreter to identify (1) areas in which the flush of growth of annual range vegetation will provide forage for livestock for a short time during the spring, and (2) areas along watercourses where salt-tolerant plant species as well as tamarisk and mesquite can be identified and their position related to soil type and availability of subsurface soil moisture. (See Color Plates 10 through 12.) Careful choice of photo dates must be made in the instances just described to take maximum advantage of phenological changes in the vegetation that are very dependent upon fluctuations in climate and rainfall. In the Bucks Lake test site, the potential for mapping gross vegetation types was discussed. Very close correlation was indicated for vegetation-type mapping using 1/30 000-scale conventional aerial photographs obtained in 1966 and the Nikon high-altitude photographs. These important cover boundaries are as recognizable on 1/950 000-scale photographs (suitably enlarged) as on conventional photography. Examples of the detection of changes in cover resulting from manipulation (logging, brush conversion, etc.) are also presented in chapter 6. Although photographs at the appropriate intervals for monitoring changes in the distribution and quantity of the winter snowpack were not obtained during 1969, an example of imagery taken on May 21, 1969, is shown and the value of the synoptic view for monitoring snow accumulation is presented.

A comparison of photointerpretability was made between Apollo 9 space photographs, taken from an altitude of approximately 120 mi with 80-mm focal length Hasselblad 70-mm cameras, and high-altitude aerial photographs, taken from an altitude of approximately 70 000 ft with 21-mm focal length Nikon 35-mm cameras. Both types of photographs, taken over agricultural areas in both California and Arizona, were studied. Our tests showed that, despite their higher spatial resolution, the aerial photographs did not significantly increase the accuracy of identification of crop types. For this reason, we believe that conclusions reached regarding the value of sequential high-altitude photographs can be applied, in general, to sequential space photographs of the quality obtained by the Apollo 9 astronauts. Especially when crop identification techniques depend upon determination of whether a field does or does not contain vegetation at particular times of the year, low-resolution photography will be quite satisfactory. These same statements are relevant to the inventory of wildland vegetation. In the Phoenix area, for example, areas covered by annual vegetation could be delineated as well on Apollo 9 photography as on HyAc panoramic photography (Color Plate 11).

For geologic mapping purposes, however, the increased resolution of the Nikon photographs was sufficient to permit more detailed mapping of geologic features. Comparisons of Apollo 9 and high-altitude photographs are made in chapter 3 that elaborate upon this point. Differences in improved interpretation for agricultural crops and geologic features suggest that decisions regarding resolution of sensing systems that provide the synoptic view (i.e., those carried aboard spacecraft and aircraft) must be made with respect to the type of natural resources to be inventoried and the quantity of data that must be extracted.

The interval necessary between successive photographic missions must also be selected in light of the particular resources being inventoried and the magnitude of change that must be determined. With respect to agricultural crops, one must decide whether within-season or between-season changes are being monitored (i.e., whether yearly or long-term patterns are being investigated). The same consideration must be made for the geologic resource. If processes are being studied that exhibit seasonal patterns (sedimentation, flooding, etc.), then photography taken one or more times each year is necessary. On the other hand, a geologic era may elapse before any significant changes can be detected in gross rock structure. Photography obtained more frequently than once each era would be repetitious for mapping rock types.

Interpretation of vegetation resources (e.g., agricultural crops in Arizona and southern California, and wildland vegetation in the Bucks Lake test site of northern California) using Infrared Ektachrome film was generally superior to that made using any of the black-and-white bands. Mapping of vegetation types was more accurately performed in the Bucks Lake test site using Infrared Ektachrome photographs than using Pan-25A or IR-89B photographs. Crop types were almost always more accurately identified on Infrared Ektachrome film than on any of the multiband photographs. However, certain vegetation types could occasionally be distinguished as well on multiband photographs as on Infrared Ektachrome film (e.g., riparian wildland vegetation, barley crops, fields of bare soil), suggesting that black-and-white photographs may be well suited for performing specific interpretation tasks.

Limited evaluation of color-enhanced images for crop identification was also made using the various systems available. Results indicate that certain enhanced images were prepared for which the accuracy of identification of particular crop types

was as great or greater than for multidate Infrared Ektachrome photographs. The potential for preparing multidate enhancements in the same manner that multiband enhancements are prepared was demonstrated using these examples.

(9) A system such as the one presented in chapter 5 for classifying range resources on space photography, or the one presented in chapter 7 for classifying timber stands and other land-use classes, can greatly improve the efficiency of a multistage sampling system that is based on space photography, aerial photography, and field observation. In the Vicksburg, Miss., area, for example, quantitative tests showed that photointerpreters could achieve an accuracy of 80 to 90 percent in making such stratifications from Apollo 9 photography. Maximum use of the "convergence of evidence principle" (as reported in ch. 3 for range resources, geologic resources, and hydrologic resources) can further improve the accuracy and usefulness of stratifications made from an interpretation of space photography.

(10) As documented in chapter 10, the various kinds of Earth resource data dealt with in this report correspond quite closely to the informational requirements of those who seek to manage Earth resources. Hence, our findings would appear to be not only of scientific interest, but also of great potential practical importance.

(11) Although, on the one hand, there are strong proponents for using aircraft rather than spacecraft for the making of Earth resource surveys, and on the other hand, those who advocate using spacecraft rather than aircraft, our findings are for the most part in support of a third view, viz, that operational Earth resource surveys of the future might best be made by means of a multistage sampling technique that employs spacecraft, aircraft, and ground observations.

(12) Because of the importance of obtaining sequential photographic coverage to aid in the inventorying of Earth resources from aerial and space photography and because cloud cover makes it very difficult to obtain sequential coverage in many geographic areas of interest, we must reiterate one important conclusion of our earlier reports: Cloud cover is likely to be the most serious deterrent to the making of operational resource surveys on photography taken from either aircraft or spacecraft.

☆ U.S. GOVERNMENT PRINTING OFFICE: 1971 O—444–666

www.ingramcontent.com/pod-product-compliance
Lightning Source LLC
Chambersburg PA
CBHW081442170526
45166CB00008B/2287